Internal Combustion Engine Technology and Applications of Biodiesel Fuel

Edited by Enhua Wang

Published in London, United Kingdom

IntechOpen

Supporting open minds since 2005

Internal Combustion Engine Technology and Applications of Biodiesel Fuel
http://dx.doi.org/10.5772/intechopen.93437
Edited by Enhua Wang

Contributors
Jayashri N. Narayanan Nair, Semakula Maroa, Freddie Liswaniso Inambao, Hongliang Luo, Zbigniew J. Jan Sroka, Claudio Marcio Santana, Jose Eduardo Mautone Barros, Ocktaeck Lim, Yanuandri Putrasari, Kenedy Aliila Greyson

First published in London, United Kingdom, 2021 by IntechOpen
IntechOpen is the global imprint of INTECHOPEN LIMITED, registered in England and Wales, registration number: 11086078, 5 Princes Gate Court, London, SW7 2QJ, United Kingdom
Printed in Croatia

British Library Cataloguing-in-Publication Data
A catalogue record for this book is available from the British Library

Additional hard and PDF copies can be obtained from orders@intechopen.com

Internal Combustion Engine Technology and Applications of Biodiesel Fuel
Edited by Enhua Wang
p. cm.
Print ISBN 978-1-83968-749-5
Online ISBN 978-1-83968-750-1
eBook (PDF) ISBN 978-1-83968-751-8

For EU product safety concerns:
IN TECH d.o.o., Prolaz Marije Krucifikse Kozulić 3, 51000 Rijeka, Croatia, info@intechopen.com or visit our website at intechopen.com.

Meet the editor

Enhua Wang has been an associate professor of Energy and Power Engineering at the Beijing Institute of Technology since 2016, teaching courses on internal combustion engines. He obtained his bachelor's and master's degrees at the Department of Automotive Engineering, Tsinghua University, Beijing, China, in 2000 and 2003, respectively. In 2013, he received his Ph.D. from the Beijing University of Technology in the field of engine combustion, conservation, and emissions control. He obtained a postdoctoral fellowship at the State Key Laboratory of Automotive Safety and Energy, Tsinghua University. In 2015, he worked at the School of Engineering, Glasgow University, Scotland, as a research associate for one year. His research interests include the organic Rankine cycle system, simulation and control of internal combustion engines, and solid oxide fuel cells.

Contents

Preface

Improving efficiency and reducing emissions of internal combustion engines, especially for transportation applications, are critical for the sustainable development of our world. Internal combustion engines with a downsized design can reduce engine size and improve fuel efficiency thanks to the turbocharging technique and electronic control system, which has great potential for passenger vehicles. The energy efficiency can be enhanced and emissions can be decreased by optimizing critical working parameters such as fuel injection and ignition timing. Meanwhile, biodiesel is produced from renewable biomass. Exploring the performance potential of biodiesel fuel provides another promising approach to replace petroleum fuel. With the help of advanced software and testing equipment, methodologies for system design and optimization of internal combustion engines are being developed. Rapid progress in computer-aided design and manufacturing technology makes it possible to design novel engine systems and reduce emissions significantly.

This book on internal combustion engines and applications of biodiesel fuel includes seven chapters on engine system design, fuel injection system analysis, ignition performance, fuel consumption estimation, application of biodiesel in engines, and more. In practice, these aspects must be considered together to maximize the potential of an internal combustion engine. The chapters are organized into two sections. The first section covers engine downsizing, fuel spray, and economic comparison. The second section deals with applications of biodiesel fuel in compression-ignition and spark-ignition engines.

A comprehensive analysis of the technology and application of the internal combustion engine is beyond the scope of the book. However, the content of this volume is useful for scientists and students to broaden their knowledge of internal combustion engine technologies and applications of biodiesel fuel.

I would like to thank IntechOpen for inviting me to be the editor of this volume, Ms. Maja Bozicevic, and the Publishing Process staff for their help in coordinating the reviews, editing, and printing of this book.

Enhua Wang
School of Mechanical Engineering,
Beijing Institute of Technology,
Beijing, China

Section 1

Internal Combustion Engine Technology

Chapter 1

Work Cycle of Internal Combustion Engine Due to Rightsizing

Zbigniew J. Sroka

Abstract

It is worth still working on the development of the internal combustion engine, because its time was not yet over. This was demonstrated by the author's review of the literature, indicating at least the perspective of 2050 the universality of the engine as the primary propulsion or support in hybrid transport units. The presented considerations may have a broader perspective, when the thermodynamic problems of a thermal machine such as an internal combustion engine are indicated. This chapter deals with the issues of changing the swept volume known as downsizing/rightsizing. An equivalent swept volume was introduced, defined by the coefficients determining changes in the cylinder diameter and the stroke of the piston. An attempt was made to find the mutual relations to the efficiency of the work cycle and engine operating parameters. The research methodology was proposed as a mix of laboratory tests and theoretical analyses, on the basis of which it was established that while maintaining the same value of the downsizing index, despite the various permissible combinations of cylinder diameter and piston stroke changes, the cycle efficiency remains unchanged. The engine operating parameters are changing, resulting from the use of support systems for rightsizing geometric changes.

Keywords: internal combustion engine, work cycle, rightsizing

1. Introduction - the essence of the research problem

It is the beginning of 2021 and internal combustion engines are not yet dead, although many people predicted their significant reduction in connection with the introduction of hybrid drive in vehicles [1]. And yet this drive still has an internal combustion engine!

When in 2007 the 2nd PTNSS Engine Congress was held in Krakow, in Poland, an international group of scientists and researchers identified three scenarios for the development of internal combustion engines:

I. short-term (until 2017): improving the design of internal combustion engines to meet ecological standards and the use of alternative fuels,

II. mid-term (2017–2037): development of hybrid systems,

III. long-term (over 30 years, i.e. over 2037): independence of transport from fossil fuels [2].

With the passage of years and the verification of predictions on the basis of real data, the need for the development of internal combustion engines was indicated indirectly in connection with the change from linear to exponential transport index passenger-kilometer, which forces the increase in the production of motor vehicles (passenger cars, trucks and busses) from the current 70 million annually to over 107 million units in 2050 [3, 4].

On August 17, 2017, Norman Mayersohn, in The New York Times magazine, in an article entitled "The Internal Combustion Engine Is Not Dead Yet", interviewed Professor John Heywood, the undisputed guru in the design and testing of internal combustion engines. Professor Heywood pointed to the presence of internal combustion engines with a significant share in 2050 – quotation: „*Definitely. John Heywood, a professor of mechanical engineering at the Massachusetts Institute of Technology, predicts that in 2050, 60 percent of light-duty vehicles will still have combustion engines, often working with electric motors in hybrid systems and largely equipped with a turbocharger. Vehicles powered purely by batteries, he estimates, will make up 15 percent of sales*" [5].

In April 2020, the virtual 41st International Vienna Motor Symposium took place (due to the COVID19 coronavirus pandemic), during which the development of internal combustion engines was discussed [6].

It was the time of the "New and Optimized Engines" session, during which Ford presented the latest solutions in the field of EcoBoost technology, emphasizing the importance of charging [7].

Toyota discussed the 1.5 Liter engine solutions from the Toyota New Global Architecture (TNGA) platform, emphasizing the importance of balance between design and application. Among other things, it was discussed: hydraulically variable valve timing, very high compression moderated by Atkinson cycle, longer bore and stroke ratio, application of multi-hole injector system to achieve "high-speed combustion", resulting more than 40% in thermal efficiency [8].

The authors of another presentation mentioned a similar meaning of the modular construction and technological platform for internal combustion engines [9].

The modularity of engines, but in relation to Diesel, was discussed during the session "New SI and CI Engines" [10], where the modular solutions of the BMW company were demonstrated.

Similar to Toyota TNGA solutions at Mercedes-Benz is FAME (Family of Modular Engines), which involves the creation of subsequent engine versions based on the M-254 engine. [11]. Everything is dedicated to the fulfillment of global CO_2 fleet targets. Quotation „..*the M 254 paves the way with regard to CO2- neutrality and air quality approaching the sustainability strategy Ambition 2039.*

The importance of the filling process both on the supercharging side and the change of the geometry of the suction system were emphasized. Attention was also paid to the reduction of friction in the piston-cylinder liner system. The summary of the whole was as follows - a quote ... *the internal combustion engine is still far from being at the end of the road!*

Environmental protection is the dominant topic in all publications. This is also the case in another study [12], where VW indicated numerous possibilities of meeting Euro 6d standards.

Subsequent studies indicate the importance of alternative fuels, with particular emphasis on hydrogen [13, 14]. The full usefulness of typical hydrogen-powered combustion engines has been demonstrated in relation to the still developed Fuel-Cell technology.

The extensive discussion is not forgotten engine applications truck [15]. Here importance is the durability of use. Considerations were conducted in the perspective of 2050!

Finally, in the general discussion, the development scenarios of combustion engines were indicated [16, 17]. In the short-term perspective, i.e. until 2030, the importance of environmental protection was emphasized, and in the longer term, i.e. until 2050, attention was additionally paid to the importance of sustainability and safe use of engines in the environment.

The above considerations have a common denominator - the world does not give up on internal combustion engines. Research centres and universities are still working on developing the design of this heat machine.

One of the development trends is the downsizing of combustion engines that has been going on for over ten years and has recently been modified towards rightsizing. This trend is not so much about reducing the displacement as it is about choosing the right size in order to achieve a balance between customer expectations for operating comfort and the manufacturer's ability to reduce fuel consumption and CO_2 emissions.

The essence of the research problem presented in this chapter is to demonstrate the existence of parameters describing the engine displacement volume, which is the dominant feature of downsizing/rightsizing, enabling the assessment of the effectiveness of changes in the IC engine operation indicators.

This means that the main research question can be formulated as follows - is it possible to replace the displacement volume of an internal combustion engine in the considerations on its work cycle, a certain equivalent volume, and can the application of the new solution be used to investigate the cause-and-effect relationships between thermodynamic parameters and internal combustion engine performance indicators?

The search for an answer to the research question is associated with the analysis of the thermodynamic work cycle of the downsized engine. The impact of changes in the swept volume and the equivalent volume on the parameters of the comparative work cycle for similar values of the downsizing index was assessed.

2. Rightsizing the internal combustion engine

Development works related to the rightsizing concept are focused primarily on increasing the specific volumetric power. These are therefore actions similar to those previously undertaken for downsizing, when reducing the displacement while maintaining or increasing the engine power per liter of displacement.

The essence of downsizing results from the power equation, which takes the form (1) [18, 19].

$$N_e = p_e V_{ss} \frac{n}{30\tau} \tag{1}$$

By changing the engine's displacement according to the rule - the volume "after" is smaller than "before", i.e. $V_{ssd} < V_{ss}$, (where: V_{ssd} is the engine swept volume after downsizing) and at the same time keeping the engine power $N_{ed} = N_e$ (where: N_{ed} - downsizing engine power), Eq. (2) is obtained

$$p_e V_{ss} \frac{n}{30\tau} = p_{ed} V_{ssd} \frac{n_d}{30\tau_d} \tag{2}$$

Indicators with the index "d" indicate downsizing data.

Assuming constancy engine speed $n_d = n$ and constancy of the number of strokes $\tau_d = \tau$, to give (3)

$$p_{ed} = p_e\left(\frac{V_{ss}}{V_{ssd}}\right) \tag{3}$$

In turn, fuel consumption expressed as the specific value (g_e) can be written as (4)

$$g_e = \frac{1}{\eta_e W_u} \tag{4}$$

where the useful efficiency η_e is expressed by the relation (5)

$$\eta_e = \frac{MRL_p p_e T_o}{\eta_v W_u p_o} \tag{5}$$

With a reasonable assumption of unchanged value beyond the engine operating after downsizing, useful efficiency becomes dependent only on the brake mean effective pressure (p_e = *BMEP*).

Brake specific fuel consumption (*BSFC* = g_e) can also be expressed by defining the actual amount of fuel burned per unit time, giving the unit of power (6).

$$g_e = \frac{G_e}{N_e} \tag{6}$$

Keeping the constant useful power after downsizing, i.e. $N_{ed} = N_e$, the following Eq. (7) is obtained

$$\frac{G_{ed}}{g_{ed}} = \frac{G_e}{g_e} \tag{7}$$

which, taking into account the close relationship between fuel consumption and the concentration of carbon dioxide in exhaust gases, will change into (8)

$$CO_{2d} = CO_2\left(\frac{g_{ed}}{g_e}\right) \tag{8}$$

The changes caused by the downsizing idea can be illustrated in the diagrams - **Figure 1**.

If, in the downsizing of the internal combustion engine, the reduction of the engine speed (*downspeeding*) is made, the effect of reducing fuel consumption and limiting carbon dioxide emissions will be enhanced. For this case, assuming the stability of the parameter values as in Eq. (9)

$$V_{ssd2} = V_{ssd1}, N_{ed2} = N_{ed1}, \tau_{d2} = \tau_{d1}, \tag{9}$$

and changing only the speed $n_{d2} < n_{d1}$, one gets (10) and (11)

$$p_{ed1} V_{ssd1} \frac{n_{d1}}{30\tau_{d1}} = p_{ed2} V_{ssd2} \frac{n_{d2}}{30\tau_{d2}} \tag{10}$$

$$p_{ed2} = p_{ed1}\left(\frac{n_{d1}}{n_{d2}}\right) \tag{11}$$

Index 1 represents the downsized base engine.

Index 2 denotes the downsized engine with changed (reduced) rotational speed.

The measure of engine modernization, both for downsizing and rightsizing, is the degree (index) of changes, which is defined in various ways [19–21]. Regardless

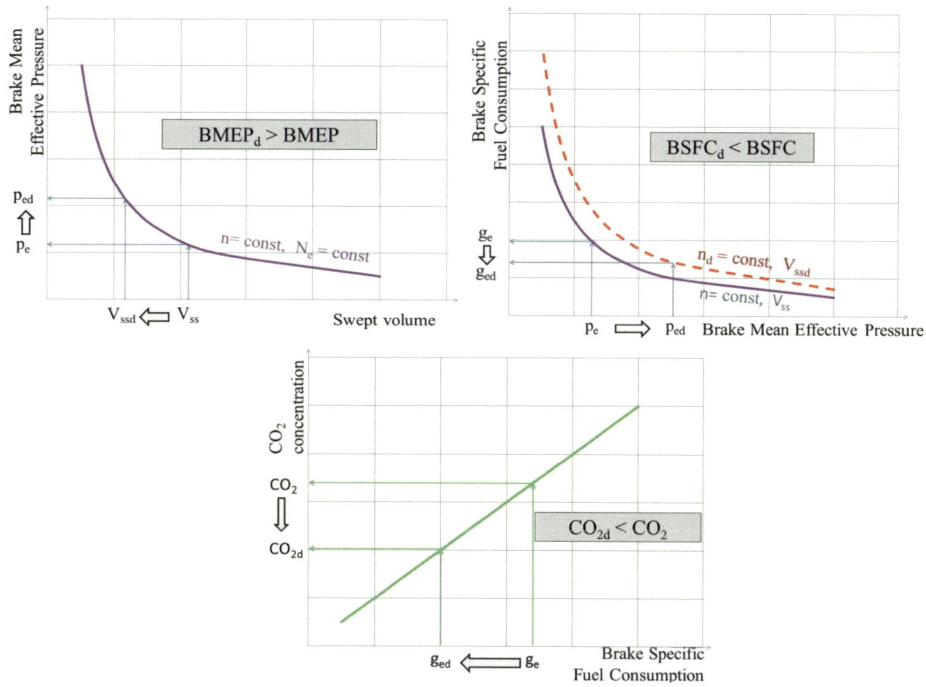

Figure 1.
The idea of downsizing.

of the definition, this indicator shows the change or degree of residue after the reduction or increase of the swept volume.

Unlike all the others, the author defined the downsizing index (W_d) based on the degrees of changes in the components describing the cylindrical combustion chamber (equivalent volume), which dominates the design of internal combustion engines [19]. According to this definition, the downsizing index can be described as in formula (12).

$$W_d = 1 - AB^2 \qquad \text{dla} \quad A = \frac{S_d}{S} \qquad B = \frac{D_d}{D} \tag{12}$$

In the graphic interpretation, theoretically and practically three forms of changes in the swept volume can be distinguished - as in **Figure 2**.

By implementing the idea of rightsizing, it is possible to obtain the same changes in the W_d index at different values of the piston stroke and the cylinder diameter, which results from the different values of the coefficients A and B (see formula 12). The downsizing/rightsizing combinations are presented in the form of a matrix of changes in the coefficients A and B - **Figure 3**. The matrix can show two volatility zones of the W_d indicator: downsizing and upsizing, important in considering rightsizing.

Having knowledge of the design of the combustion chamber and the crank system in the commonly accepted geometric relationships between the cylinder diameter and the stroke [18, 22], as well as based on the actual relationships of these parameters determined on the basis of the engines from the Engine of the Year competition over the years 1999–2019 [19, 20] it was possible to determine the real ranges of variability of the ratio of the cylinder diameter to the piston stroke, which is from 0.77 to 1.30, which results in the value of the W_d index in the range of minus −1.20 on the upsizing side and plus +0.51 in the case of downsizing.

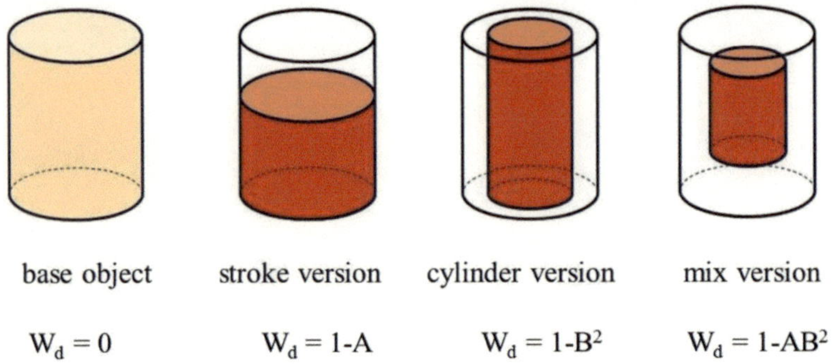

Figure 2.
Forms of downsizing [19].

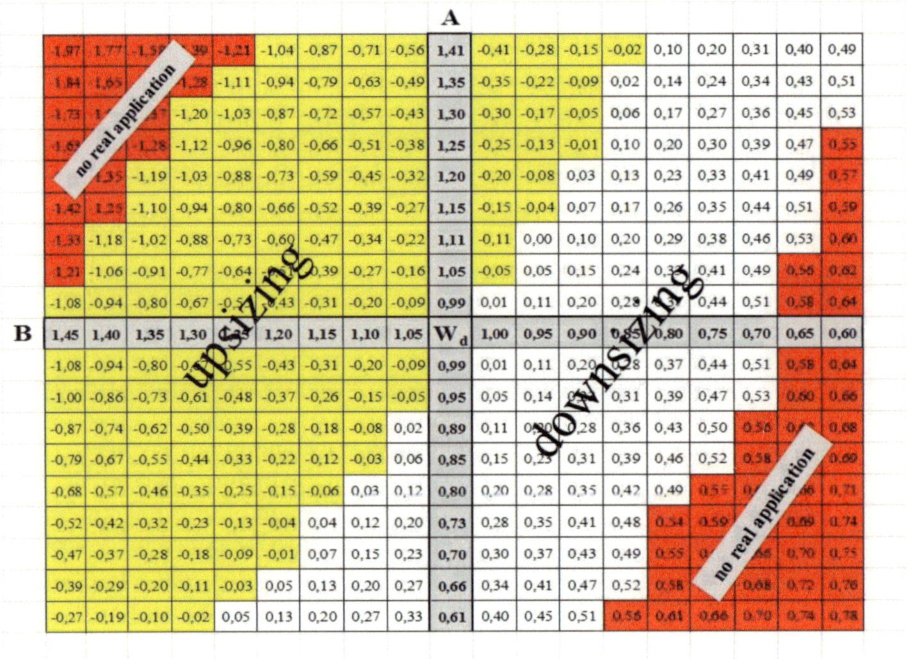

Figure 3.
Matrix of changes to the downsizing/rightsizing index according to various combinations of coefficients A and B (according to formula 12).

In order to maintain the operational parameters of the internal combustion engine, while reducing its stroke volume, it is necessary to implement new or intensify the existing functions performed by individual structural and functional systems in the engine. Among them, an important place is occupied by: direct fuel injection, charging, variable valve timing, variable compression ratio. And the whole thing is controlled by electronics [23–25].

The idea of direct fuel injection developed differently in the two different engine types (diesel and gasoline). It has been used almost always in diesel engines, but the implementation of the Common Rail system by the Denso/Toyota corporation played a special role. It happened in 1995, although the idea was known as early as 1916 (Vickers company) [26]. However, at that time, there was no technology of obtaining high pressure, atomization of fuel drops and the possibility of multiple fuel injection in one cycle [27]. Today, as a result of this, fuel consumption is reduced and the emission of harmful exhaust components is significantly reduced due to the lower temperature in the combustion chamber. Additionally, a lower noise level is achieved, which significantly improves the comfort of operation [28].

On the other hand, the implementation of direct gasoline injection in spark ignition engines resulted in greater positive effects in the economic and ecological balance of engine development. The first attempts to inject gasoline directly into the combustion chamber were carried out by Jonas Hesselman in 1925, but only the solution proposed by Mitsubishi in 1996 brought success in development. This solution is known as GDI - Gasoline Direct Injection [29]. Gasoline injection, carried out in at least two phases during the intake and compression stroke, allows for stratified combustion, including combustion of very poor mixtures (50: 1 versus stoichiometric - conventional 14.7: 1), which in turn helps to increase the compression ratio without knocking effect. The use of a special combustion chamber geometry in the piston crown and thus achieving a load swirl increases engine power with a simultaneous reduction in fuel consumption. The disadvantage of this system is, unfortunately, the increase in nitrogen oxides emissions, which means that the engine must be equipped with a reducing catalyst and exhaust gas recirculation system. Of great importance in the implementation of GDI is control, including adaptive systems [30]. The use of direct injection fits very well into the architecture of the engine covered by downsizing/rightsizing because it directly complements the power loss resulting from changes in geometry.

Another downsizing/rightsizing support system is the charging, the presence of which is essential for proper cylinder filling. As early as 1885, Gottlieb Daimler noticed the need for charging to increase the filling level in his patent about the need to increase the air pressure above atmospheric at the beginning of each cycle [18]. Then came the concept of recycling the energy wasted with the exhaust gas outlet and in 1916 Auguste Reteau built the first turbocharger. For many years, the concept of a single turbocharger functioned until the appearance of the Honeywell turbocharger, where, due to the limited response time to changes in engine load on a common axle, two compressor wheels appeared next to one turbine. The engine with such a system works more efficiently, especially in the lower engine speed (rpm) and load ranges. In the following years, various solutions began to appear, including variable VNT (Variable Nozzle Turbine) settings. An interesting solution is the system of two turbochargers working in parallel, which replace one large one. Thanks to this solution, the turbochargers are smaller (in line with the downsizing idea), which results in less heat loss to the atmosphere.

There are also combinations of mechanical, electric and traditional charging. [31–33]. Supercharging is the simplest form of supporting the downsizing/rightsizing engine, both in terms of power loss and by creating conditions for burning poor mixtures to meet ecological requirements.

The improvement of volumetric efficiency is also achieved by the application of variable valve timing systems. The variable valve timing system ensures that the angles and times of opening and closing the valves are matched to the current load and engine speed.

There are many variable valve timing systems which undergo successive design transformations and take different names depending on the manufacturer [34]. The first variable valve timing system appeared in 1981 on Alfa Romeo engines, but it was only the introduction of electronic control in 1989 by Honda that allowed the development of this design known as VTEC (Variable Valve Timing and lift Electronic Control), and in the latest version i- VTEC (i - intelligent system that works ahead).

In contrast, the VarioCam system, designed by Porsche in 1992, altered the position of the valves by changing the tension in the chain connecting the intake and exhaust camshafts. Today the system is developed and also offers valve lift capability. Another example is the Valvetronic system from BMW with full control of the intake valve lift, which significantly reduces flow losses and the reaction time to load changes is reduced to a minimum.

Yet another example in this field is Ford's TI-VCT (Twin Independent - Variable Camshaft Timing) system of independent inlet and outlet valve operation, whose main advantage over other systems is better cylinder filling and scavenging the combustion chamber.

The variable valve timing system is a good complement to the downsizing/rightsizing technique by being able to reduce flow losses due to smaller valve dimensions and by ensuring that the combustion chamber is properly filled to maintain or increase engine efficiency.

When supercharging spark-ignition engines, there may be a risk of spontaneous combustion, which is inherently undesirable. In order to prevent this, the compression ratio should be lowered, which in turn determines the pressure in the combustion chamber, and this affects the engine power throughout its entire operating range. The solution to this problem is a system with a variable compression ratio.

The principle of operation of the variable compression ratio system - VCR is associated with a change in the volume of the compression chamber with the change of load. There are several technical solutions to this issue. One of them is the change of stroke in the crank mechanism (Multi Cycle Engine 5, implemented by Peugeot).

Another way is the angular displacement of the cylinder head offered by SAAB (SVC system - Saab Variable Compression). Yet another solution is the dynamic movement of the entire crank system (Cortina VC - Variable Compression). The GoEngine solution is structurally interesting as it provides a change in the compression ratio in the range from 8: 1 to 18: 1. A significant advantage of this system is the possibility of a significant (up to 20%) extension of the expansion stroke in relation to the compression stroke, which provides better conditions for burning the fuel dose, generates more favorable pressure distribution on the piston crown and lowers the exhaust gas temperature. A variable compression ratio system, by varying the cylinder volume, can be considered one of the forms of dynamic downsizing/rightsizing, not as a support system.

From engineering practice, there are a number of examples of the development of the downsizing/rightsizing idea. We can even mention the engines installed in Ford or Volkswagen vehicles.

The Ford's engine with a displacement of 2.3 dm^3 V6 was reduced to 2.0 dm^3 and 1.6 dm^3, to finally reach the spectacular 0.999 dm^3 EcoBoost - **Figure 4**. Some people consider the engine with a displacement of 5.0 dm^3 Coyote to be the progenitor of all downsizing/rightsizing changes. This makes changes a kind of cascade of actions.

In turn, the Volkswagen engines changed the displacement from 2.8 dm^3 or 2.0 dm^3 to 1.8 dm^3, and then to 1.4 dm^3, fulfilling the downsizing assumption, and with sustainable development (rightsizing) the 1.4 dm^3 engine was replaced with 1.5 dm^3.

On a large scale, the trend of changing the displacement volume is well represented by the engines considered in the international competition Engine of The Year, which since 1999 has been organized by the magazine Engine Technology International - UK & International Press [35]. The winning engines in all categories show a clear trend of change in displacement over the years. It is expressed by an increase in the specific power and a decrease in carbon dioxide emissions, which increase with a decrease in the stroke volume - **Figure 5**.

In automotive practice, internal combustion engines designed in the downsizing and rightsizing technique can be found in cars with a whole package of

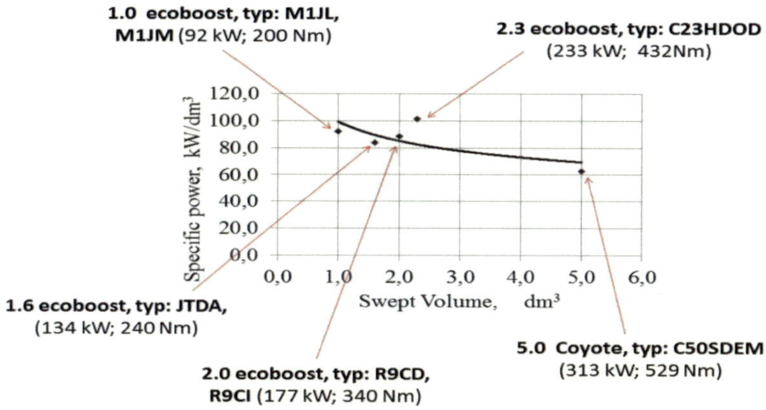

Figure 4.
Rightsizing on the example of Ford's engines.

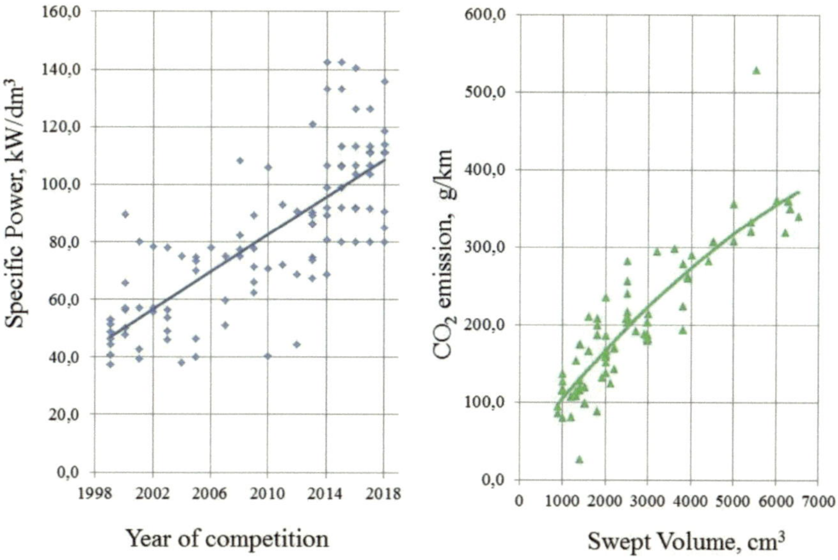

Figure 5.
The specific power of combustion engines together with the carbon dioxide emissions of the "winners" in the engine of the year competition in all categories.

pro-ecological solutions and are included in marketing names, for example: EcoBoost/Econetic (Ford) or Blue Motion (Volkswagen) [2].

3. Efficiency of the generalized engine work cycle in terms of rightsizing - research methodology

In the combustion chamber of a reciprocating internal combustion engine, the fuel mixed with air creates a working medium that undergoes thermodynamic changes, related, among other things, to the volume of the combustion space. These changes are repeatable, although their magnitude depends on the current operating conditions of the engine. The occurring transformations create the engine work cycle, mathematically described in various ways [36–38]. In a generalized form, corresponding to all known theories of internal combustion engines, the work cycle

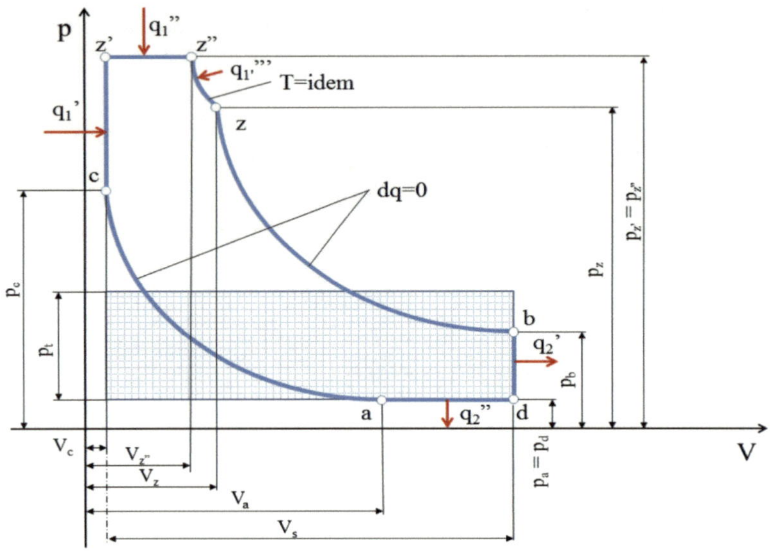

Figure 6.
Generalized thermodynamic work cycle of a four-stroke internal combustion engine [36].

can be described by the efficiency (ηt) as per formula (13) and expressed graphically as in **Figure 6**.

$$\eta_t = 1 - \frac{\frac{\lambda_p \rho_p \varepsilon_s^{\kappa-1}}{\delta^{\kappa-1}} + \kappa(\rho' - 1) - \rho'}{\varepsilon_s^{\kappa-1}\left\{\lambda_p\left[\kappa\rho_p - (\kappa-1)\left(1+\rho_p \ln \rho_T\right)\right] - 1\right\}} \quad (13)$$

The individual dimensionless quantities appearing in formula (13) are described in accordance with **Figure 6** [19].

- degree of pressure increase during isochoric heat transfer

$$\lambda_p = \frac{p_{z'}}{p_c} = \frac{p_{z''}}{p_c} \quad (14)$$

- degree of expansion during isobaric heat transfer

$$\rho_p = \frac{V_{z''}}{V_{z'}} = \frac{V_{z''}}{V_c} \quad (15)$$

- effective compression ratio

$$\varepsilon_s = \frac{V_a}{V_c} \quad (16)$$

- isentropic exponent

$$\kappa = \frac{c_p}{c_v} \quad (17)$$

- degree of another expansion process

$$\delta = \frac{V_b}{V_z} \quad (18)$$

- degree of pre-compression when heat is drained at constant pressure

$$\rho' = \frac{V_d}{V_a} = \frac{V_b}{V_a} \tag{19}$$

- degree of expansion during isothermal heat transfer

$$\rho_T = \frac{V_z}{V_{z''}} \tag{20}$$

- geometric compression ratio

$$\varepsilon = \frac{V_b}{V_c} = \frac{V_d}{V_c} \tag{21}$$

By introducing the quantities expressed by the formulas (14)–(21) into the formula (13), one can obtain relationships that emphasize the changes in various volumes, which can be used to describe the changes caused by downsizing (22)

$$\eta_t = 1 - \frac{\frac{\lambda_p \left(\frac{V_{z''}}{V_c}\right)\left(\frac{V_a}{V_c}\right)^{\kappa-1}}{\left(\frac{V_b}{V_z}\right)^{\kappa-1}} + \kappa\left[\left(\frac{V_b}{V_a}\right) - 1\right] - \frac{V_b}{V_a}}{\left(\frac{V_a}{V_c}\right)^{\kappa-1}\left\{\lambda_p\left[\kappa\left(\frac{V_{z''}}{V_c}\right) - (\kappa-1)\left(1 + \left(\frac{V_{z''}}{V_c}\right)\ln\left(\frac{V_z}{V_{z''}}\right)\right)\right] - 1\right\}} \tag{22}$$

λp, Va, Vz, Vz'', κ are components resulting from the properties of the fuel used and the logistics of the combustion process, while V_b and V_c are design parameters of the internal combustion engine related to the combustion space, and therefore related to the rightsizing operation.

The introduction to formula (22) of the variables A and B from formula (12) gives a full picture of changes in thermodynamic transformations in the theoretical cycle of the downsizing/rightsizing engine. When assessing the effectiveness of applying the rightsizing idea, three cases can be considered:

1. all the considered components are subject to change, that is: the displacement volume $Vsd \neq Vs$ together with the compression volume $V_{cd} \neq V_c$ and the comprcssion ratio $\varepsilon_d \neq \varepsilon$ (23)

$$\eta_{td} = 1 - \frac{\frac{\lambda_{pd}\left(\frac{V_{z''d}}{V_c\frac{(\varepsilon-1)}{(\varepsilon_d-1)}AB^2}\right)\left(\frac{V_{ad}}{V_c\frac{(\varepsilon-1)}{(\varepsilon_d-1)}AB^2}\right)^{\kappa_d-1}}{\left(\frac{\left(V_s+V_c\frac{(\varepsilon-1)}{(\varepsilon_d-1)}\right)AB^2}{V_{zd}}\right)^{\kappa_d-1}} + \kappa_d\left[\left(\frac{\left(V_s+V_c\frac{(\varepsilon-1)}{(\varepsilon_d-1)}\right)AB^2}{V_{ad}}\right) - 1\right] - \frac{\left(V_s+V_c\frac{(\varepsilon-1)}{(\varepsilon_d-1)}\right)AB^2}{V_{ad}}}{\left(\frac{V_{ad}}{V_c\frac{(\varepsilon-1)}{(\varepsilon_d-1)}AB^2}\right)^{\kappa_d-1}\left\{\lambda_{pd}\left[\kappa_d\left(\frac{V_{z''d}}{V_c\frac{(\varepsilon-1)}{(\varepsilon_d-1)}AB^2}\right) - (\kappa_d-1)\left(1 + \left(\frac{V_{z''d}}{V_c\frac{(\varepsilon-1)}{(\varepsilon_d-1)}AB^2}\right)\ln\left(\frac{V_{zd}}{V_{z''d}}\right)\right)\right] - 1\right\}} \tag{23}$$

If we assume that the selection of the compression ratio for the downsizing/rightsizing engine will be made on the basis of experimental data, e.g. by comparing the compression ratio values of the engines included in the *Engine of the Year* competition, then for typical examples the relationship between ε and ε_d was identified [19] (16).

$$\varepsilon_d = 0.547\varepsilon + 4.239 \tag{24}$$

It means the possibility of introducing a new coefficient (C), expressed by the relation (25).

$$\frac{\varepsilon_d - 1}{\varepsilon - 1} = C \tag{25}$$

After taking into account the dependence (25), the formula describing the theoretical efficiency of the engine work cycle takes the form (26).

$$\eta_{td} = 1 - \frac{\dfrac{\lambda_{pd}\left(\frac{V_{z''d}}{V_c\frac{AB^2}{C}}\right)\left(\frac{V_{ad}}{V_c\frac{AB^2}{C}}\right)^{\kappa_d-1}}{\left(\frac{\left(V_s+V_c\frac{1}{C}\right)AB^2}{V_{zd}}\right)^{\kappa_d-1}} + \kappa_d\left[\left(\frac{\left(V_s+V_c\frac{1}{C}\right)AB^2}{V_{ad}}\right) - 1\right] - \frac{\left(V_s+V_c\frac{1}{C}\right)AB^2}{V_{ad}}}{\left(\frac{V_{ad}}{V_c\frac{AB^2}{C}}\right)^{\kappa_d-1}\left\{\lambda_{pd}\left[\kappa_d\left(\frac{V_{z''d}}{V_c\frac{AB^2}{C}}\right) - (\kappa_d-1)\left(1+\left(\frac{V_{z''d}}{V_c\frac{AB^2}{C}}\right)\ln\left(\frac{V_{zd}}{V_{z''d}}\right)\right)\right] - 1\right\}} \tag{26}$$

2. the following are subject to change: the swept volume $V_{sd} \neq V_s$ and the compression volume $V_{cd} \neq V_c$ without changing the compression ratio $\varepsilon_d = \varepsilon$ (27)

$$\eta_{td} = 1 - \frac{\dfrac{\lambda_{pd}\left(\frac{V_{z''d}}{V_cAB^2}\right)\left(\frac{V_{ad}}{V_cAB^2}\right)^{\kappa_d-1}}{\left(\frac{(V_s+V_c)AB^2}{V_{zd}}\right)^{\kappa_d-1}} + \kappa_d\left[\left(\frac{(V_s+V_c)AB^2}{V_{ad}}\right) - 1\right] - \frac{(V_s+V_c)AB^2}{V_{ad}}}{\left(\frac{V_{ad}}{V_cAB^2}\right)^{\kappa_d-1}\left\{\lambda_{pd}\left[\kappa_d\left(\frac{V_{z''d}}{V_cAB^2}\right) - (\kappa_d-1)\left(1+\left(\frac{V_{z''d}}{V_cAB^2}\right)\ln\left(\frac{V_{zd}}{V_{z''d}}\right)\right)\right] - 1\right\}} \tag{27}$$

3. thc third case is the change of the swept volume $V_{sd} \neq V_s$ and the compression ratio $\varepsilon_d \neq \varepsilon$ without changing the compression space $V_{cd} = V_c$ (28)

$$\eta_{td} = 1 - \frac{\dfrac{\lambda_{pd}\left(\frac{V_{z''d}}{V_c}\right)\left(\frac{V_{ad}}{V_c}\right)^{\kappa_d-1}}{\left(\frac{V_sAB^2+V_c}{V_{zd}}\right)^{\kappa_d-1}} + \kappa_d\left[\left(\frac{V_sAB^2+V_c}{V_{ad}}\right) - 1\right] - \frac{V_sAB^2+V_c}{V_{ad}}}{\left(\frac{V_{ad}}{V_c}\right)^{\kappa_d-1}\left\{\lambda_{pd}\left[\kappa_d\left(\frac{V_{z''d}}{V_c}\right) - (\kappa_d-1)\left(1+\left(\frac{V_{z''d}}{V_c}\right)\ln\left(\frac{V_{zd}}{V_{z''d}}\right)\right)\right] - 1\right\}} \tag{28}$$

In the test evaluation methodology, the real values of A and B coefficient pairs are introduced from the matrix described in **Figure 2**. This way, changes in the thermodynamic work cycle efficiency can be calculated. The rest of the data was taken from research on the 1.4 TSI, 1.5 TFSI, 1.8 T and 2.0 TDI engines, which are an example of a link in the downsizing/rightsizing chain of Volkswagen engines.

The study covered an extreme case of changes, i.e. changes in both the swept volume, compression and compression ratio (formula 23).

To evaluate the research problem, theoretical and experimental data from the tests of the VW 1.4 TSI internal combustion engine carried out at the Department of Vehicle Engineering of the Wroclaw University of Science and Technology - **Figure 7**. The next data were used from the tests on the chassis dynamometer of vehicles equipped with 1.8 T and 2.0 TDi engines - **Figure 8**.

The research data constituting the boundary conditions for the evaluation of the 1.5 TFSI engine were obtained from the literature [40].

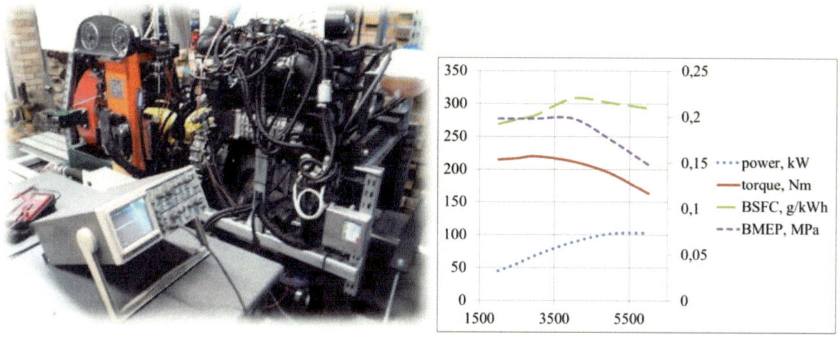

Figure 7.
VW 1.4 dm^3 engine on the test stand [39].

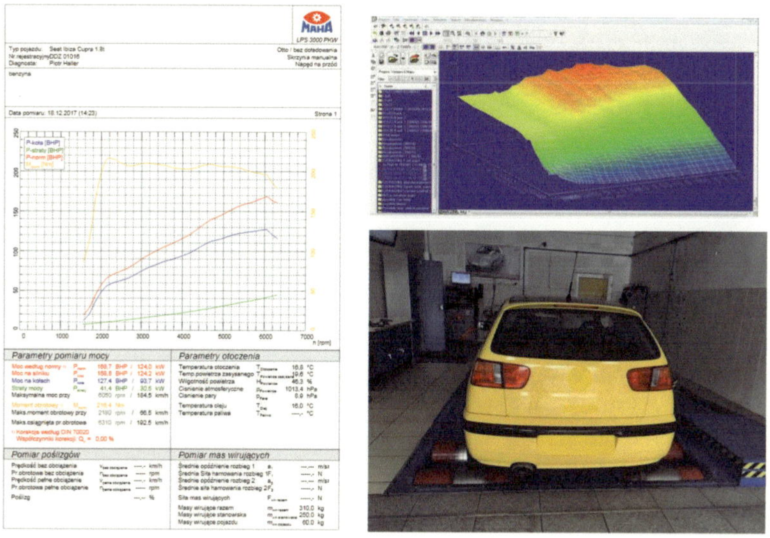

Figure 8.
Tested vehicle with a 1.8 T engine on a chassis dynamometer.

The downsizing/rightsizing indexes according to formula (12) for the cascade of changes of the swept volume are as follows:

- 2.0 dm^3 na 1.8 dm^3 $W_d = 0.09$
- 2.0 dm^3 na 1.5 dm^3 $W_d = 0.25$
- 2.0 dm^3 na 1.4 dm^3 $W_d = 0.29$

For each case, apart from the factory version, theoretical changes related to the behavior of the W_d index with different coefficients A and B, taken from the matrix of changes - **Figure 2**, were considered.

In this way, a package of variables was obtained and analyzed - **Table 1**.

It is worth noting that in the case of the 1.4 dm^3 engine, in which downsizing/rightsizing according to the form "cylinder version" (A = 1) was intended, the rule of mutual relation of the diameter and stroke of the piston, which should be in the range (0.77–1.30) - as discussed above. Hence the decision to change the relation to the closest unity to A = 0.97.

Engines	S	A	S_d	D	B	D_d	W_d	D/S	D_d/S_d	Remarks
dm^3	mm	—	mm	mm	—	mm	—	—	—	
2.0	92.80	—	—	82.50	—	—	0	0.89	—	base
1.8		0.91	84.10		1.00	82.50	0.090		0.98	Factory-1.8_1
1.8		1.00	92.80		0.955	78.80	0.088		0.85	Test-1.8_2
1.8		0.97	90.00		0.97	80.00	0.087		0.89	Test-1.8_3
1.5		0.93	85.9		0.90	74.50	0.247		0.87	Factory-1.5_1
1.5		0.75	69.60		1.00	82.5	0.250		1.19	Test-1.5_2
1.5		1.00	92.80		0.87	71.50	0.243		0.77	Test-1.5_3
1.4		0.82	75.60		0.93	76.50	0.291		1.01	Factory-1.4_1
1.4		1.00	92.80		0.84	69.30	0.294		0.75	Does not meet $0.77 \leq D/S \leq 1.30$
1.4		0.97	90.00		0.855	70.50	0.291		0.78	Test-1.4_2
1.4		0.71	65.00		1.00	82.50	0.290		1.25	Test-1.4-3

Table 1.
The stroke and cylinder diameter values and the corresponding coefficients define the downsizing/rightsizing area.

The values of the A and B coefficients taken for the assessment filled successive values of the downsizing/rightsizing W_d index, ensuring their invariability within a given cylinder volume. The remaining data, filling the form of the formula for the efficiency of the comparative cycle with the equivalent volume (formula 23) and enabling the evaluation of the engine performance indicators, were obtained from the above-mentioned laboratory tests.

4. Discussion of the results

Typical operating indicators of engine work were assessed together with parameters of the thermodynamic cycle, including the efficiency of the generalized work cycle. The obtained data are presented in the form of relative changes, i.e. as a percentage of the data for the base engine 2.0 dm^3 - **Tables 2–4**.

The data contained in **Table 2** refer to the 1.8 dm^3 engine and confirm the correctness of the downsizing idea due to the reduction in fuel consumption by an average of 5%. Thanks to the support systems with supercharging at the forefront and control of the combustion process, even an increase in power of nearly 14% compared to the 2.0 dm^3 unit was achieved.

Greater efficiency was obtained both on the theoretical and useful side. The differences between the efficiency changes η_e and η_t are due to exhaust losses and cooling.

It is worth emphasizing that the change of the coefficients A and B, in such a way that the downsizing index W_d is maintained, did not cause significant differences in the values of all examined parameters and fell within the limits of statistical significance.

The data contained in **Table 3** refer to the 1.5 dm^3 engine and confirm the correctness of the downsizing concept due to the reduction in fuel consumption by an average of nearly 20%. The proposal to reduce the stroke volume by about 25% is close to aggressive downsizing.

Parameter	Manufacturer-2.0	Manufacturer-1.8_1	Test-1.8_2	Test-1.8_3
		% 1.8/2.0	% 1.8/2.0	% 1.8/2.0
ε	10.5	−4.7	−4.7	−4.7
rpm	6000	−8.3	−8.3	−8.3
n_1	1.35	−1.5	+0.4	−1.0
n_2	1.19	0	−0.2	0
T_{max}, K	2706	−0.7	+0.1	−0.5
η_v	0.92	+29.1	+28.9	+29.1
BMEP	1.11	+36.8	+35.9	+36.0
BSFC, g/kWh	264	−5.6	−5.2	−5.0
N_e, kW	110	+13.6	+13.7	+13.7
η_e	0.32	+6.0	+5.5	+5.3
η_t	0.45	+0.7	+0.4	+0.4

Table 2.
*Values of selected engine operating parameters 1.8 dm^3 in relation to 2.0 dm^3 at different values of the downsizing/rightsizing coefficients A and B (**Table 1**).*

Parameter	Manufacturer-2.0	Manufacturer-1.5_1	Test-1.5_2	Test-1.5_3
		% 1.5/2.0	% 1.5/2.0	% 1.5/2.0
ε	10.5	+19.1	+19.1	+19.1
rpm	6000	−16.7	−16.7	−16.7
n_1	1.35	+9.0	+9.0	+9.7
n_2	1.19	−2.5	−2.5	−2.5
T_{max}, K	2706	+6.6	+6.6	+7.5
η_v	0.92	+26.6	+26.6	+29.7
BMEP	1.11	+54.3	+56.7	+59.4
BSFC, g/kWh	264	−19.7	−20.9	−19.5
N_e, kW	110	+2.9	+2.1	+0.22
η_e	0.32	+24.5	+26.4	+24.2
η_t	0.45	+6.9	+7.2	+6.8

Table 3.
*Values of selected engine operating parameters 1.5 dm^3 in relation to 2.0 dm^3 at different values of the downsizing/rightsizing coefficients A and B (**Table 1**).*

The approx. 27% increase in volumetric efficiency is due to the boost system and variable valve timing set. The engine power was retained with the seemingly reasonable boost, which resulted in a greater than the 1.8dm^3, but less than 1.4dm^3 increase in BMEP. In the group of tested engines, it is the only engine in which the *downspeeding* concept was applied, changing the maximum value of the engine speed from 6000 to 5000 rpm. There was no significant increase in temperature in the maximum work cycle. Keeping the downsizing/rightsizing index W_d at the level of 0.25, it was shown that the change of coefficients A and B does not cause differentiation of the theoretical work cycle efficiency.

Parameter	Manufacturer-2.0	Manufacturer-1.4_1	Test-1.4_2	Test-1.4_3
		% 1.4/2.0	% 1.4/2.0	% 1.4/2.0
ε	10.5	−4.7	−4.7	−4.7
rpm	6000	0	0	0
n_1	1.35	+7.5	+7.4	+9.6
n_2	1.19	−3.4	−3.4	−2.5
T_{max}, K	2706	+5.0	+5.0	+6.5
η_v	0.92	+38.4	+39.0	+37.8
BMEP	1.11	+62.2	+60.4	+62.0
BSFC, g/kWh	264	−13.7	−12.4	−14.0
N_e, kW	110	+13.6	+13.6	+13.5
η_e	0.32	+15.9	+14.2	+13.5
η_t	0.45	+2.7	+2.7	+2.6

Table 4.
*Values of selected engine operating parameters 1.4 dm^3 in relation to 2.0 dm^3 at different values of the downsizing/rightsizing coefficients A and B (**Table 1**).*

The data in **Table 4** refer to the 1.4 dm^3 engine and indicate an aggressive downsizing of up to 30%. The expected effect was achieved, i.e. the specific fuel consumption was reduced by an average of 13%, which obviously translates into a reduction in carbon dioxide emissions to the atmosphere. The implementation of support systems for geometric changes resulted in a significant increase in BMEP by over 60%, which may result in a reduction in the durability of engine parts, especially in the area of the piston and crank system.

The differences between the efficiency changes η_e and η_t are due to losses in the exhaust and cooling systems.

The change of coefficients A and B does not significantly affect, and even the differences in values are insignificant, on the tested parameters.

From the point of view of rightsizing, it should be noted a clear relationship between the cycle efficiency and the necessary change in the stroke volume, i.e. one that will correspond to a sustainable approach to design by meeting customer needs and at the same time fulfilling the manufacturer's capabilities.

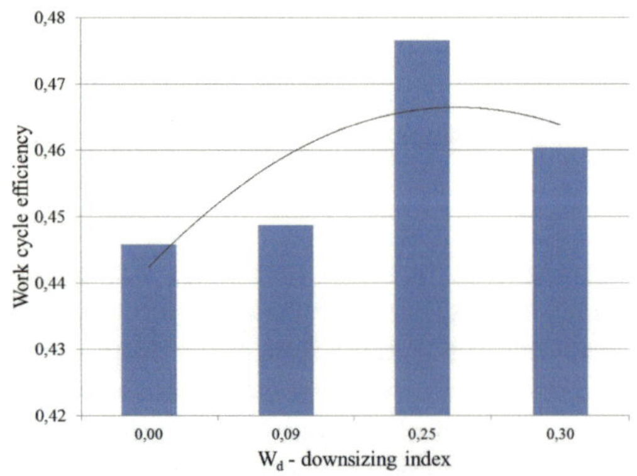

Figure 9.
Changes in the efficiency of the engine work cycle in relation to the downsizing index.

Figure 9 presents the relationship between the cycle efficiency and the downsizing index, which shows that the reduction of the swept volume will be effective up to a certain limit. For the analyzed case, the implementation of an engine with a volume of 1.5 dm^3 instead of 1.4 dm^3 is an example of this.

5. Summary

The problem of changing the displacement of the internal combustion engine is known as downsizing, but recently it has been undergoing a transformation towards rightsizing. It is the result of a new approach to the design and operation process, which assumes balancing the customer's requirements and the manufacturer's capabilities, all in a specific environment, e.g. constantly tightening ecological standards. Anyway, the ecological aspect is the most desirable criterion for assessing the rightsizing concept, which is expressed in the pursuit of reducing fuel consumption and the resulting reduction in carbon dioxide emissions, all for the correct selection of the engine's displacement volume.

A research problem was defined, which is the assessment of the influence of the differences in geometrical changes of the piston stroke and the cylinder diameter, while maintaining the same value of the downsizing index, on the efficiency of the engine work cycle.

To fulfill the aim of the research, the relationship describing the theoretical efficiency of the general reference cycle was modified, in which instead of the stroke volume, substitute coefficients defining changes in the value of the piston's stroke (A) and the cylinder diameter (B) were revealed.

A spectacular case of changes was adopted for the considerations, assuming a change in the swept volume and the accompanying changes in the compression space and compression ratio. The necessary data for the analysis was obtained from laboratory tests and from the literature. There have been repeatedly estimated intermediate quantities defining the efficiency of the engine supported by the obtained engine operation parameters.

The analysis of the results shows that the efficiency of the internal combustion engine cycle is stable, regardless of the A and B coefficients, which determine the geometric changes of the engine displacement volume.

The presence of the limit value of the downsizing/rightsizing index was also demonstrated, at which the highest level of positive change in the circulation efficiency is achieved, corresponding to the sustainability requirements.

In the next steps, the research work must be targeted at detailed on-road exhaust gas toxicity testing of downsizing/rightsizing engines. Due to the significant load on the engine structure, it will also be important to pay attention to the issues of material engineering and tribological processes related to the downsizing/rightsizing concept.

Acknowledgements

The works were carried out in the GEO-3EM research complex of the Wroclaw University of Science and Technology, in the laboratories of the Department of Automotive Engineering.

This research was funded by Wroclaw University of Science and Technology, grant number MPK 9100560000/8201003902.

Nomenclature

A	coefficient of change of piston stroke
B	coefficient of change of cylinder diameter
BMEP = p_e	brake mean effective pressure
BSFC = g_e	brake specific fuel consumption
D	cylinder diameter - input state
D_d	cylinder diameter in the downsized engine
Ge	fuel consumption per hour
L_p	moles index for ambient air
MR	universal gas constant
N_e	useful power
p_o	ambient pressure
rpm = n	engine revolution
S	stroke of the piston - input state
S_d	stroke of the piston in the downsized engine
T_o	ambient temperature
V_{ss}	engine swept volume
W_d	downsizing/rightsizing index
W_u	fuel calorific value
δ	degree of another expansion process
ε	geometric compression ratio
ε_s	effective compression ratio
η_t	theoretical efficiency of the work cycle
κ	isentropic exponent
λ_p	degree of pressure increase during isochoric heat transfer
ρ'	degree of pre-compression when heat is drained at constant pressure
ρ_p	degree of expansion during isobaric heat transfer
ρ_T	degree of expansion during isothermal heat transfer
τ	stroke index (number of strokes)

Author details

Zbigniew J. Sroka
Wroclaw University of Science and Technology, Faculty of Mechanical Engineering, Department of Automotive Engineering, Wroclaw, Poland

*Address all correspondence to: zbigniew.sroka@pwr.edu.pl

References

[1] Warnecke W., Lueke W., Clarke L., Louis J., Kempsel S., Fuels of the Future. Proceedings of 27th International Vienna Motor Symposium, Vienna 2006.

[2] Wisłocki K., Wolański P., Ecker H., Lundqvist U., Pearson R.J., Hartland J., Biernat K., Czerwinski J., Wyszyński M., Powertrain development from the perspective of panel discussions at the second International PTNSS Congress, Combustion engines 2/2007 (129), 38–53

[3] Lenz H.P., 30 International Vienna Motor Symposium. 7–8 May 2009 – Report on the occasion of the International Congress PTNSS on Combustion Engines 2009 in Opole, Combustion Engines 2/ 2009 (137), 150–154.

[4] Walsch M.P., Global trends in motor vehicle pollution control: a 2011 update – part 3, Combustion Engines 4/2011 (167), 98–103.

[5] Mayersohn N. The Internal Combustion Engine Is Not Dead Yet, The New York Times Magazine, 17th August 2017

[6] Geringer B., Lenz H.P.,41st International Vienna Motor Symposium, 22–24 April 2020, Reports

[7] Ruhland H., Wirth M., Friedfeld R., Linsel J., Weber C., Krämer F., Ford Werke GmbH, Cologne; Abkenar F., Ford Motor Company, Dearborn, USA: EcoBoost 500: Taking Award Winning Technology to the Next Level, Reports 41st International Vienna Motor Symposium, 22–24 April 2020,

[8] Kitadani H., Kaneda R., Mizoguchi S., Shinohara Y., Takeuchi J., Toyota Motor Corporation, Toyota, Japan: The New 1.5 Liter Gasoline Engine from the TNGA Series, Reports 41st International Vienna Motor Symposium, 22–24 April 2020,

[9] Song D., Hycet e-Chuang, Great Wall Motor, Hebei, China; W. Happenhofer, Great Wall Motor, Hebei, China: 1.5T High Thermal Efficiency Modular Engine Platform, Reports 41st International Vienna Motor Symposium, 22–24 April 2020,

[10] Steinparzer F., Hiemesch D., Kranawetter E., Salmansberger M., Stütz W., BMW Motoren GmbH, Steyr: The Technical Concept of the New BMW 6-Cylinder 2nd Generation Modular Diesel Engines, Reports 41st International Vienna Motor Symposium, 22–24 April 2020,

[11] Dr. T. Schell, Mercedes-Benz AG, Stuttgart: M254 – the Future of the 4-Cylinder Gasoline Engine, Reports 41st International Vienna Motor Symposium, 22–24 April 2020,

[12] Helbing C., Köhne M., Kassel T., Wietholt B., Krause A., Lohre L., Gerhardt N., Eiglmeier C., Volkswagen AG, Wolfsburg: Volkswagen's TDI-Engines for Euro 6d – Clean Efficiency for Modern Mobility, Reports 41st International Vienna Motor Symposium, 22–24 April 2020,

[13] Schwieberdingen; Univ.-Prof. Dr. H. Eichlseder, Dr. P. Grabner, Dr. K. Schaffer, Graz University of Technology: H2 ICE for Future Passenger Cars and Light Commercial Vehicles, Reports 41st International Vienna Motor Symposium, 22–24 April 2020,

[14] Korn T., KEYOU GmbH, Unterschleißheim: The Most Efficient Way for CO2 Reduction: the New Generation of Hydrogen Internal Combustion Engines, Reports 41st International Vienna Motor Symposium, 22–24 April 2020,

[15] Lozanovski A., Geß A., University of Stuttgart; Dipl.-Ing. O. Dingel,

Dipl.-Ing. (FH) T. Semper, IAV GmbH, Chemnitz: Technical Evaluation and Life Cycle Assessment of Potential Long Haul Heavy Duty Vehicles for the Year 2050, Reports 41st International Vienna Motor Symposium, 22–24 April 2020,

[16] Pischinger S. - RWTH Aachen University; van der Put D., Heuser P. - FEV Group GmbH, Aachen; Lindemann B., Müther M., Schönen M. - FEV Europe GmbH, Aachen: Efficient Commercial Powertrains – How to Achieve a 30% GHG Reduction in 2030, Reports 41st International Vienna Motor Symposium, 22–24 April 2020,

[17] Hartung S., Member of the Board of Management, Chairman Business Sector Mobility Solutions, Robert Bosch GmbH, Stuttgart: Powertrains of the Future – Sustainable, Safe, Exciting, Reports 41st International Vienna Motor Symposium, 22–24 April 2020,

[18] Heywood J.B., Internal Combustion Engine Fundamentals, McGraw HiU International Editions 1989.

[19] Sroka Z.J., Wybrane zagadnienia teorii tłokowych silni-ków spalinowych w aspekcie zmian objetości skokowej, Oficyna Wydawnicza Politechniki Wrocławskiej, 2013

[20] Fraser A.D.J., How Low can we go? Challenges and opportunities of Engine Downsizing to reduce CO2 Emissions, Seminar Proceedings IMechE, London, 9 February 2011, 1–9.

[21] Pielecha I., Cieślik W., Borowski P., et al. Reduc-tion of the number of cylinders in internal combustion engines – contemporary trends in downsizing. Combustion Engines. 2014, ISSN 2300–9896, 159(4), 12–25.

[22] Pischinger S, Verbrennungskraftmaschinen I, RWTH Aachen, Aachen 2011.

[23] Fraser N., Bassett M., Extreme Engine Downsizing with a single Turbochanger – 100 kW/l and 30 bar BMEP, Seminar Proceedings IMechE, London, February 2011, 31–45.

[24] Jenteges M., van der Weem D., et al. Optimized Activation of a Downsizing Concept with Electrical Boost, MTZ 04/ 2006, Vol. 67.

[25] King J., Application of Synergistic Technologies to Achive High Levels of Gasoline Engine Downsizing, Seminar Proceedings IMechE, London, 9 February 2011, 59–72.

[26] Fisher C.H., Carburation, Vol. III, Spark-Iquition Engines Fuel Injection Systems, Chapman & Hall, London 1966.

[27] Lejda K., Woś P., Fuel Injection in Automotive Engineering – simulation of combustion process in direct injection diesel engine based on fuel injection characteristics, InTech., 2012.

[28] Lejda K., Injection systems of high-speed diesel engines and development trends, Combustion Engines 4/2005 (123), 19–30.

[29] King J., Application of Synergistic Technologies to Achive High Levels of Gasoline Engine Downsizing, Seminar Proceedings IMechE, London, 9 February 2011, 59–72.

[30] Wendeker M., Adaptacyjne sterowanie wtryskiem benzyny w silniku, Państwowe Wydawnictwa Naukowe, Warszawa 2000.

[31] Kammeyer J., Natkaniec C., Seume J.R., Influence of tip-qap losses on the stage efficiency of downsizing turbochanger turbines. Proceedings of 9th International Conference on Turbochangers and Turbocharging (IMechE) 10.1243/ 17547164C0012010023, London, 19–20 May 2010, 293–306

[32] Lake T., Stokes J., Murphy R., Downsized DI Gasoline Engines for Low CO2. Seminar Proceedings IMechE, Fuel Economy and Engine Downsizing, London, 13 May 2004, 49–55.

[33] Wijetunge R., Criddle M., Dixon J., Morris G., Retaining Driveability in Aggressively Downsized Diesel Engines. Seminar Proceedings IMechE, Fuel Economy and Engine Downsizing, London, 13 May 2004, 41–47.

[34] Mitianiec W., Bac G., Camless hydraulic valve timing system in combustion engines, Combustion Engines 3/2011 (146), 28–37.

[35] www. http://www.ukimediaevents.com/engineoftheyear/

[36] Ambrozik A., Wybrane zagadnienia procesów cieplnych w tłokowych silnikach spalinowych, Wydawnictwo Politechniki Świętokrzyskiej, Kielce 2003.

[37] Ambrozik A., Analiza cykli pracy czterosuwowych silników spalinowych, Wydawnictwo Politechniki Świętokrzyskiej, Kielce 2010.

[38] Blair G.P., Design and Simulation of Four-Stroke Engine, Society of Automotive Engineers, Warrendale 1999.

[39] Sroka Z.J., Dworaczyński M., Assessment of thermodynamic cycle of internal combustion engine in terms of rightsizing, Combustion Engines, 178 (3), 2019

[40] Hordecki J. Volkswagen Golf 1.5TSI – odwórcenie trendu, https://www/auto-swiat.pl (13 Feb, 2017)

Chapter 2

Experimental Investigations on Fuel Spray and Impingement for Gasoline Direct Injection Engines

Hongliang Luo

Abstract

Spray-wall impingement is a widespread phenomenon applied in many fields, including spray-wall cooling system, spray coating process and fuel spray and atomization in internal combustion engines. In direct-injection spark ignition (DISI), it is difficult to avoid the fuel film on the piston head and cylinder surfaces. The wet wall caused by impingement affects the air-fuel mixture formation process, which finally influence the subsequent combustion efficiency and performance. Therefore, the fuel spray and impingement under gasoline engine-like conditions were characterized. Mie scattering technique was applied to visualize the spray evolution and impingement processes in a high-pressure and high-temperature constant chamber. Meanwhile, the adhered fuel film on the wall was measured by refractive index matching (RIM) under non-evaporation and evaporation conditions considering the effects of different injection pressures, ambient pressures and ambient temperatures. Additionally, the fuel film formation and evaporation evolution models were proposed with the help of these mechanisms.

Keywords: fuel spray, impingement, fuel film, non-evaporation and evaporation

1. Introduction

Generally, compared to port fuel injection (PFI) engine, direct-injection spark ignition (DISI) engines as a greatly potential alternative become more and more widely used for their significant advantages [1, 2]. However, owing to the short impingement distance and high injection pressure, spray impingement on the cylinder wall and piston head is quite difficult to avoid. The impingement affects the fuel-air mixture prior to combustion, which is a possible source for unburned hydrocarbon (UHC) and particulate matters (PM) [3]. Tanaka et al. [4] showed the relationship between the deterioration ratio of engine particle number (PN) emissions and fuel film volume, indicating that PN emissions increases with the increasing of fuel film on the wall.

As we known, when fuel immediately out of the nozzle hole, the spray breakup occurs. Two processes of fuel breakup can be involved. The first breakup is also called the primary breakup, leading to large droplets and liquid ligaments near the nozzle to form the dense spray [5, 6]. Under the high injection pressure condition, the cavitation and turbulence are generated from the injector holes, which should be the main reason for this mechanism. Then the following breakup process is named as secondary breakup, indicating these existing droplets break up into smaller ones owing to aerodynamic forces caused by the relative velocity between droplets and surrounding ambient gas. During the secondary break-up, more and more liquid droplets are formed and move downstream of the spray. Although considerable researches were done for the spray and atomization mechanisms, the spray and impingement are less discussed, let along the fuel film formation after impingement. By considering the competitions from the pure electric vehicle (EV) and concerns on environmental pollution, it is urgent to understand the interaction of liquid droplet and piston wall thoroughly to improve the spray atomization in engine work process.

The target of this study is to investigate the characteristics of gasoline spray impingement and fuel film formation experimentally. The specific objectives of this study are shown as follows:

- Analyze the impinging spray evolution characteristics which belongs to the single-hole nozzle under the non-evaporation and evaporation conditions.

- Model the fuel film formation to provide an insightful understanding of the interaction between spray and wall.

In this study, Mie Scattering technique was implemented to obtain the observations of gasoline spray emerging from single hole injector under different conditions. RIM technique was adopted to analyze the formation process of fuel liquid on the wall qualitatively and quantitatively under both non-evaporation and evaporation conditions.

2. Experimental apparatus and measurement methods

As shown in **Figure 1**, a quartz glass (Sigma Koki, DFSQ1-50CO2) was used as the impingement wall with a diameter at 50 mm and thickness at 2 mm.

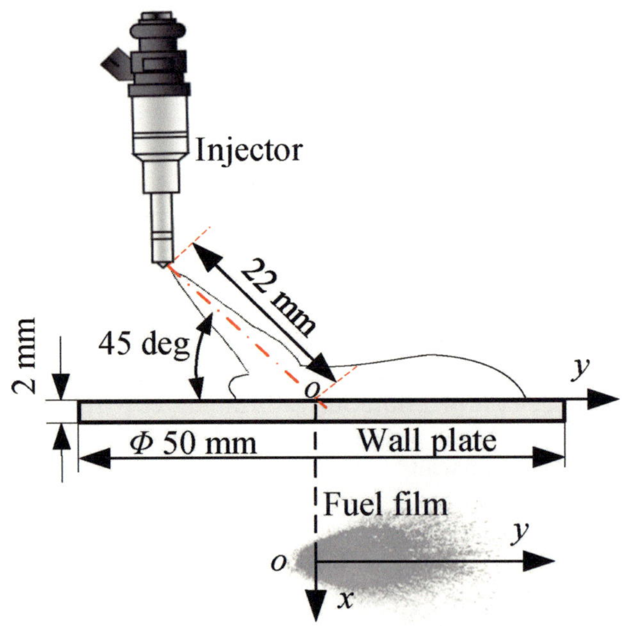

Figure 1.
Schematic of injector and flat wall.

The coordinate system is established and defined, and the intersection point o of the nozzle center axis and the flat wall is decided as the impingement point. The positive y axis is along the lateral direction of the spray after impingement, and the positive x axis is pointing out of the figure. The impingement angle was 45 deg. and the impingement distance was 22 mm from the nozzle exit to the impingement point of the flat wall along the spray axis. Moreover, the surface roughness was measured at Ra 7.0 μm by a portable high-performance surface roughness and waviness measuring instrument (Kosaka Laboratory Ltd., SE300).

The experiment was performed in a constant high-pressure chamber filled with nitrogen gas. A single hole mini-sac injector was used with length at 0.65 mm and hole diameter at 0.155 mm. The test conditions are listed in **Table 1**. Toluene was selected as a substitute for gasoline. The fuel temperature (before injection) was regulated by a cooling system to maintain it at room temperature. The injection pressure changes among 10, 20 and 30 MPa, resulting in the different duration at 2.9, 2.1, and 1.7 ms for fuel injection, which satisfies the constant injection mass at 4.0 mg by considering the real injection mass in each hole. In order to study the influence of ambient pressure, pressures between 0.15 and 0.74 MPa were tested at evaporation condition (T_{amb} = 433 K). Meanwhile, the equivalent non-evaporating conditions (P_{amb} = 0.1 MPa, T_{amb} = 298 K and P_{amb} = 0.5 MPa, T_{amb} = 298 K) were determined by maintaining the ambient density at 1.95 and 5.95 kg/m^3, respectively. One thing should be pointed out that the saturated temperatures (T_{sat}) of toluene under P_{amb} = 0.15 MPa is 398 K, and it is cleat to see that $T_{sat} < T_{amb}$. While under P_{amb} = 0.74 MPa, T_{sat} is 472 K, higher than T_{amb}.

2.1 Mie scattering method

In order to observe the spray development process, the Mie scatting experiment was performed, and the specific experimental apparatus are shown in **Figure 2**. A high-speed video camera (Photron FASTCAM SA-Z) was utilized to observe the spray with a frame rate at 20,000 frames per second (fps) and a frame size at 512 × 512 pixels. A xenon lamp (Ushio SX-131 UID501XAMQ) was set in a direction perpendicular to the camera to illuminate the spray, and it was placed in the same plane with the camera.

The spray tip penetration (S) and impinging spray height (H_i) are widely used to investigate the characteristics of the spray-wall impingement [7, 8]. These values were experimentally obtained from raw images by determining the edge of

	Non-evaporation conditions	Evaporation conditions
Test Fuel	Toluene	
Fuel Temperature	298 K	
Injection Mass	4.0 mg	
Ambient Gas	Nitrogen	
Injection Pressure	10, 20, 30 MPa	
Injection Duration	2.9, 2.1, 1.7 ms	
Ambient Temperature	298 K	433 K
Ambient Pressure	0.1 and 0.5 MPa	0.15 and 0.74 MPa
Ambient Density	1.95 and 5.95 kg/m^3	1.95 and 5.95 kg/m^3

Table 1.
Test conditions.

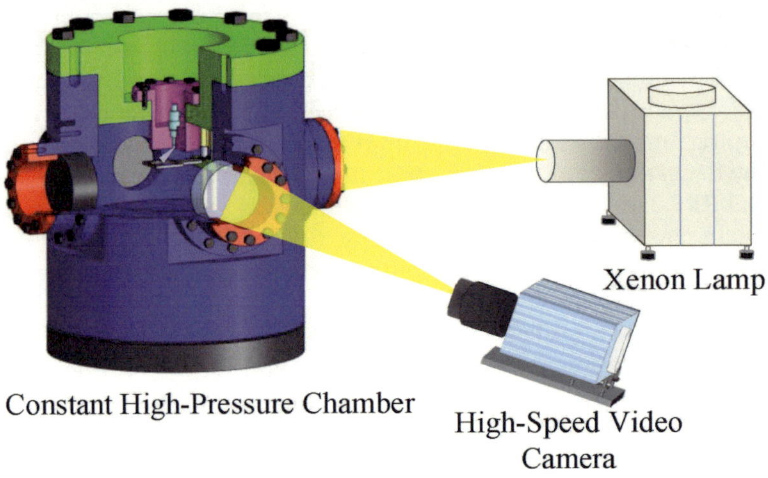

Figure 2.
Experimental apparatus.

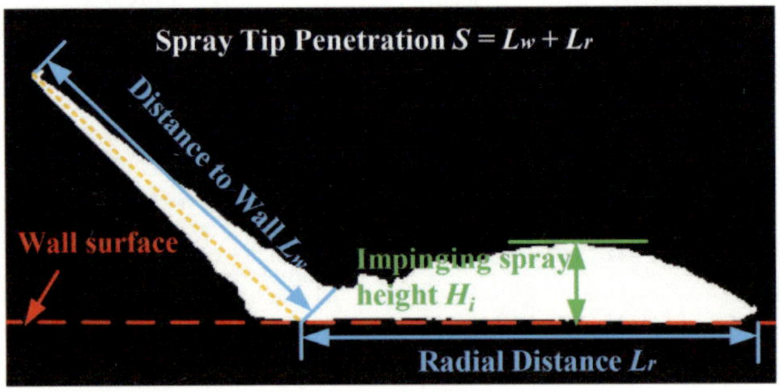

Figure 3.
Experimental extraction of S and H_i.

the impinging spray using inhouse code created in MATLAB software as shown in **Figure 3**. The dotted line represents the wall surface. *S* and H_i are defined. *S* is defined as the distance from the nozzle exit to the spray tip. Generally, before impingement, *S* is just the distance from the nozzle exit to the spray tip. However, after wall impingement, *S* is defined as the sum of the distance to wall (L_w) and the radial distance (L_r) [9]. H_i is the maximum distance from the wall surface to the edge of the impinging spray. All the results were calculated five times under each specific set of experimental conditions, and the average values were presented.

2.2 RIM method

The RIM experimental equipment in the current study is shown in **Figure 4**. Some differences can be found in the optical system with Mie scattering. The xenon lamp was placed at the side window to emit continuous light with an incident angle at 15 deg. Under the flat wall, a reflection mirror was positioned directly. Fuel film images were observed by a high-speed video camera through this mirror. Further, the high-speed video camera was set at a frame rate of 10,000 fps and at a frame size of 512 × 512 pixels.

RIM method is applied to measure the fuel film thickness. The image processing is shown in **Figure 5**. First, the image without fuel film named as "dry image" is acquired. After fuel adhering on the wall, it is subtracted by the "dry image"

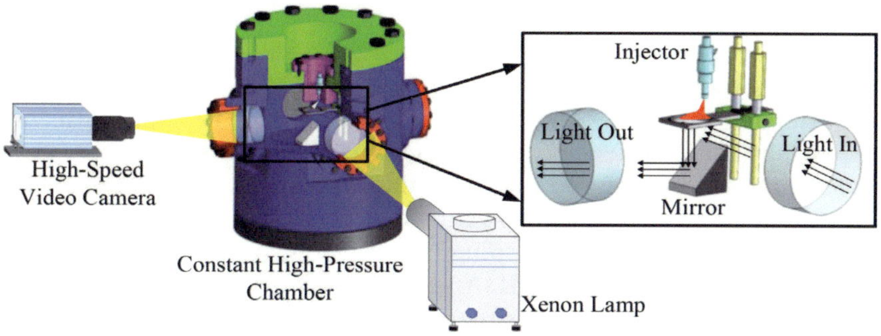

Figure 4.
Experimental setup for fuel film measurement.

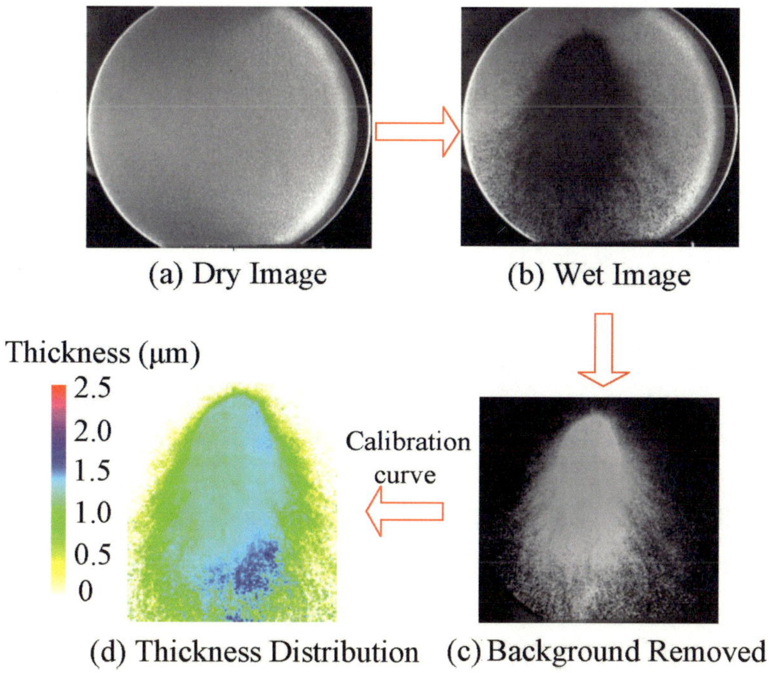

Figure 5.
Image processing.

to obtain the only adhered film image. Using the calibration result, the thickness distribution can be obtained. With the scale (0.106 mm/pixel) got from the observation, the film area is calculated by integrating the available pixels. While, for the mass, the thickness can be added up if larger than 0.1 μm. Then the film mass can be calculated through the scale and density of toluene (867 kg/m^3). Additional details about RIM method and calibration can be seen in the previous publications [10–13].

3. Impinging spray under evaporation conditions

3.1 Effect of temperature

To discuss the spray characteristics under different ambient temperatures, the images of spray and fuel film at 1.5 ms and 4.0 ms after start of injection (ASOI) under T_{amb} = 298 K are presented in **Figure 6(a)**. The sprays are displayed as binary images acquired from the Mie scattering experiment, and the fuel films are shown in pseudo-color images obtained from the RIM experiment. During injection, the fuel film is incomplete and invalid owing to the mie scattering effect caused by the illumination of the droplets above the wall. While, the fuel film becomes complete and thus available after injection. Therefore, all the results related to the adhesion are shown and discussed after fuel injection. Besides, it is clear to see that the impinging spray height at 4.0 ms ASOI is a little larger than that at 1.5 ms ASOI. And more droplets accumulated at the downstream should be one possible reason for it. Furthermore, the fuel film at impingement region (called Region I) is thicker than that at the periphery region (called Region II) and the thickness decreases along y direction. This is due to different mechanisms in the formation of fuel film, which will be explained detailly in the following section.

The images of spray and fuel film at 1.5 ms and 4.0 ms ASOI under T_{amb} = 433 K can be seen in **Figure 6(b)**. Under condition of high ambient temperature, the impinging spray height is shorter than that under T_{amb} = 298 K, owing to the evaporation of fuel droplets. This phenomenon is more obvious at 4.0 ms AOSI. Moreover, the spray area is smaller in contrast to that observed under T_{amb} = 298 K, which can be also explained by the evaporation of droplets during spray. By comparing **Figure 6(a)** and **(b)**, it is evident that evaporation is more significant at the periphery than that observed on Region I.

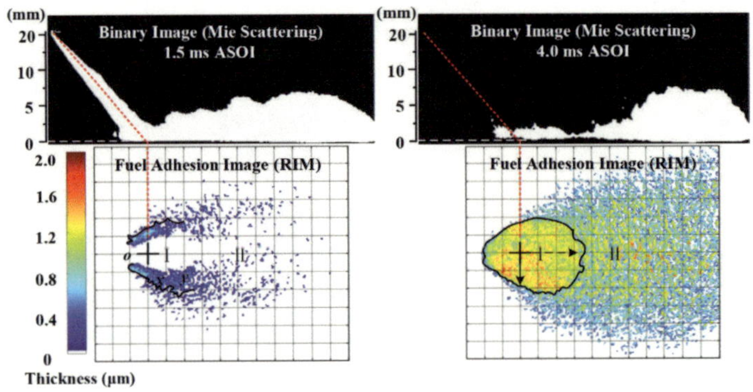

(a)

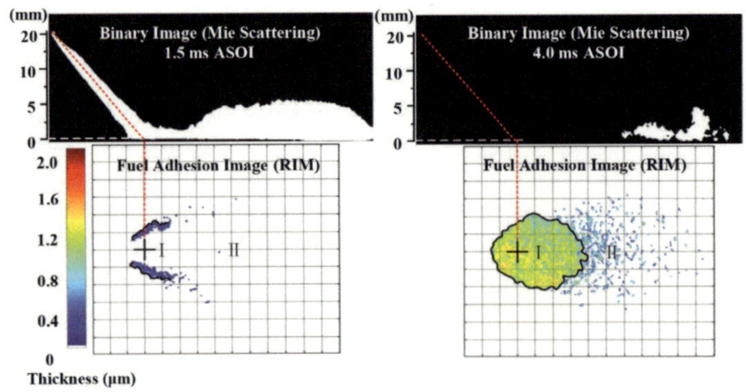

(b)

Figure 6.
Spray impingement and fuel film during and after injection. (a) T_{amb} *= 298 K. (b)* T_{amb} *= 433 K.*

For clarifying the effect of ambient temperature on fuel film thickness distribution along different lines, which were clearly defined to express them concisely. As illustrated in **Figure 7**, the coordinate system is created by setting the impingement point as the origin o. As fuel film is almost symmetric along the y axis, thickness distribution can be only discussed at different y lines at y = −2.5, 2.5, 7.5, and 12.5 mm, and the thickness at these lines was described as $H_{-2.5}$, $H_{2.5}$, $H_{7.5}$, and $H_{12.5}$, respectively. The results of $H_{-2.5}$, $H_{2.5}$, $H_{7.5}$, and $H_{12.5}$ at 5 and 50 ms ASOI under T_{amb} = 298 K and 433 K are depicted in **Figure 8**, respectively. The horizontal axis is from −15 to 15 mm, and the vertical axis denotes fuel film thickness.

The results at 5 ms ASOI are shown in **Figure 8(a)**. The thickness under T_{amb} = 298 K is larger than that under T_{amb} = 433 K due to non-evaporation. For all cases except of $H_{12.5}$, thickness distribution can be divided into three parts as "initially increases, remains constant, and finally decreases". For the constant value, $H_{-2.5}$ and $H_{2.5}$ under 298 K are similar to those under 433 K, but $H_{7.5}$ under 433 K is slightly smaller than that under 298 K. Then as the constant values under 298 K and 433 K are similar, this region is defined as Region I. While for the other region, whose value is variable is defined as Region II. The boundary to separate Regions I and II can be detected through the constant value in thickness using our in-house code made in the MATLAB, which is marked in **Figure 7(a)**. And it is depicted that the Region I under 298 K is almost the same as that under 433 K. A similar observation can be seen in **Figure 7(b)**, although difference can be seen in $H_{-2.5}$, as $H_{-2.5}$ under 433 K evaporates for a quite long time at 50 ms ASOI, leading to a sharp reduction. Regions I and II are presented in **Figure 7(b)** by using the same criterion. It is worth noting that Region I at 5 ms and 50 ms ASOI are similar as well.

Overall, the mechanisms of fuel film formation are illustrated in **Figure 9** by applying the divided Regions I and II as mentioned above.

Region I also can be named as primary impingement region. The spray impinges on the wall directly, leading to some fuel sticking on it to form the film.

Region II also can be named as secondary impingement region. After fuel spray impingement, most of the fuel splashes off it. The splashing droplets are re-deposited on the wall to form the film because of coalescence and air force.

During injection: When fuel is ejected from the nozzle, liquid fuel breaks up into ligaments and large droplets. Through the spray propagation, the ligaments and large droplets break up into small ones owing to the interaction between the liquid fuel and ambient gas. In the case of T_{amb} = 298 K, after the primary impacting, some liquid fuel deposits on the wall to form Region I, and other fuel splashes off the wall. However, the splashed droplets may collide and coalesce with others, causing a change in velocity. Finally, these droplets may redeposit on the wall to form Region II. Furthermore, under high temperature conditions, less fuel can be seen on Region II,

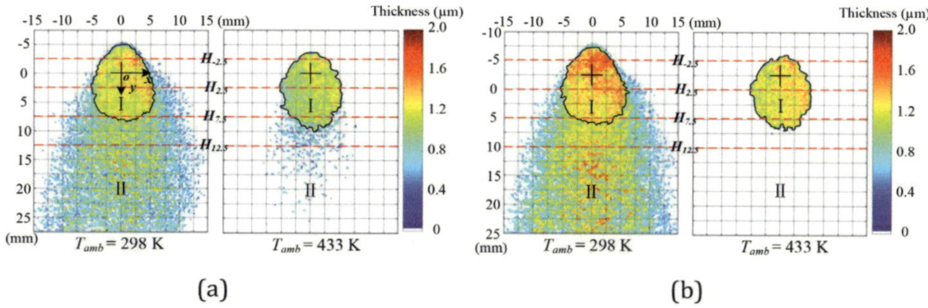

Figure 7.
Fuel film under different ambient temperatures. (a) 5 ms ASOI (b) 50 ms ASOI.

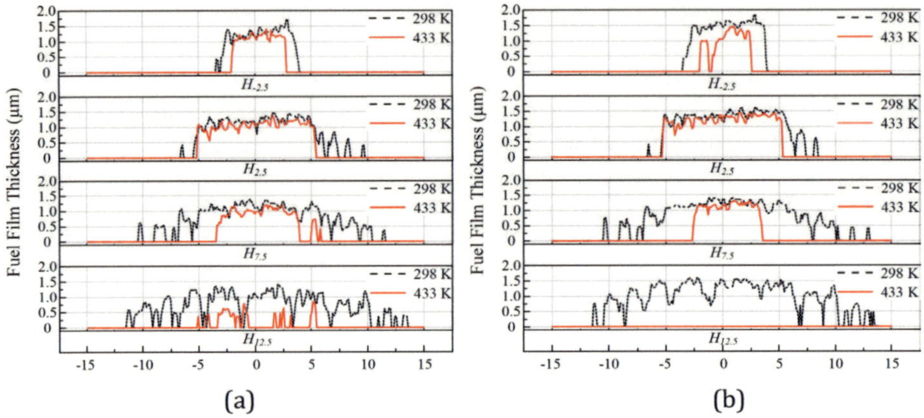

Figure 8.
Fuel film thickness along different lines. (a) 5 ms ASOI (b) 50 ms ASOI.

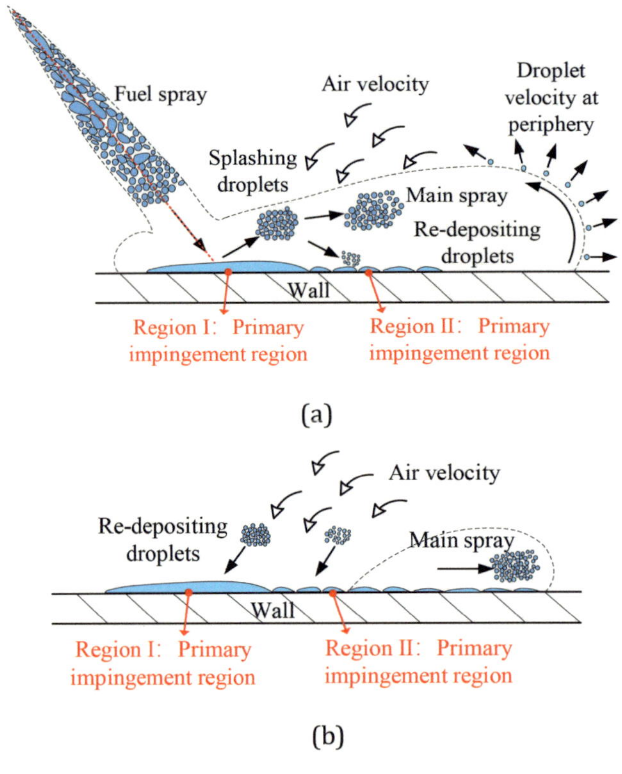

Figure 9.
Mechanisms of fuel film formation in different regions. (a) During injection. (b) After injection.

and the decreases in impinging droplets and fuel film evaporation should be the reasons for it.

After injection: Although no fuel spray can be seen, some tiny droplets are still in the air above the wall. Under T_{amb} = 298 K, the entrainment air changes the velocity of the splashing droplets. As a result, these droplets may re-impact on the wall to form Regions I and II. Under T_{amb} = 298 K, the same phenomena can be expected. But with considering the evaporation effect, the splashing droplets evaporate so quickly that few droplets can redeposit on the wall to from Region II. The fuel film on Region I is mainly formed by the initial impingement, less effect can be seen on Region I, causing only periphery of film evaporating, which leads to fuel film on Region I almost the same as that under room temperature.

To summarize, the fuel film on Region I is caused by the direct impinging spray, affected less by the high ambient temperature. However, film on Region II is mainly caused by the redeposition of the splashing droplets. Under high temperature condition, the splashing droplets evaporate easily before reattaching the wall. Even some can redeposit to form film on the wall, it evaporates quickly due to the strong heat and mass transfer. As a result, fuel film on Region I is similar under T_{amb} = 298 K and 433 K, but varies greatly on Region II, as shown in **Figure 7**.

3.2 Effect of injection pressure

Figure 10 shows *S* under different injection pressures varying among 10, 20 and 30 MPa. Results at T_{amb} = 298 and 433 K are depicted by solid and open data. The horizontal axis represents time after start of fuel injection, and the vertical axis is *S*. The impingement distance is shown by dotted line. Owing to the fuel evaporation, *S* at T_{amb} = 433 K is slightly lower than that at T_{amb} = 298 K before impingement. Nevertheless, the difference becomes obvious after impingement, which can be attribute to the impingement facilitating fuel breakup and atomization. Moreover, larger difference in *S* can be found under P_{inj} = 10 MPa. And the longer time for spray propagation and evaporation may be one possible explanation for it. Although high injection promotes better atomization, the accelerated fuel spray shortens the spray development time, which might be another reason for explaining larger difference existing at low injection pressure.

H_i can be seen in **Figure 11**. The dotted line represents the end of injection (EOI) with the injection duration depicted. The impingement timing can be advanced under high injection pressure because of high momentum. At EOI, H_i at P_{inj} = 30 MPa is larger than that at 10 MPa under both evaporation and non-evaporation condition. The larger Weber number induced by high injection pressure results in more splashing droplets should be the reason for it. Although high injection pressure promotes better atomization, leading to fast evaporation of fuel, the stronger splashing phenomenon cannot be hindered. Furthermore, after EOI at T_{amb} = 433 K, as no fuel supply, H_i under P_{inj} = 30 MPa decreases more sharply

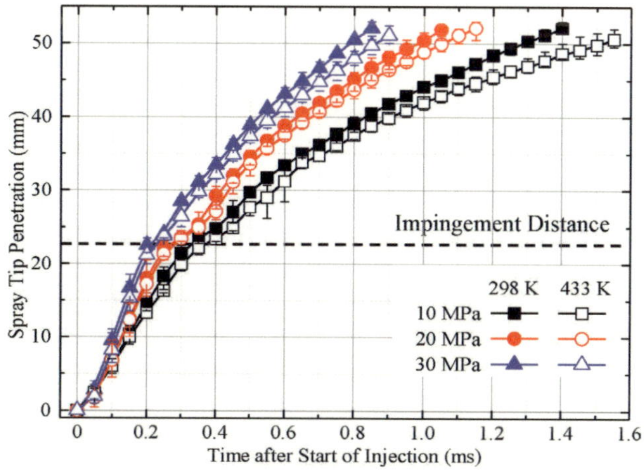

Figure 10.
Spray tip penetration under different injection pressures.

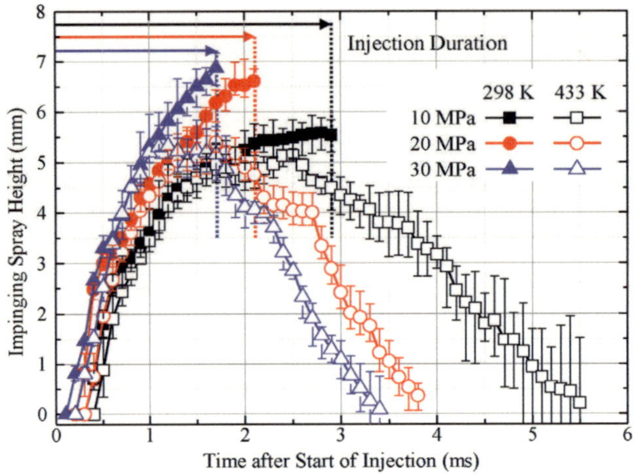

Figure 11.
Impinging spray height under different injection pressures.

than that under 10 MPa. And the better atomization and fast evaporation should be responsible for it.

In order to discuss the effect of injection pressure on the fuel film evolution, different timing at 5, 20, and 40 ms ASOI are selected to show at different injection pressures under T_{amb} = 298 K and 433 K, as shown **Figure 12**. The cross symbol is used to present the impingement point, and color bar is applied to show the thickness from 0 to 2.0 μm. It is important to see the maximum thickness is only 2 μm, and some reasons can be involved to explain it. Firstly, when liquid droplets impacting on the wall, various behaviors can be concluded such as "stick", "spread" and "splash", which are determined by the Weber number of the incident droplets [14]. Therefore, most droplets after impacting could splash off the wall, left few adhering on the wall. Secondly, the rough surface of the wall also promotes the droplets breakup and splashing of droplets after impingement [15]. Thirdly, the formed fuel film may be destroyed by incoming droplets and carried away by the air flow, leading to the maximum thickness at 2 μm. In addition, Ding et al. [16] and Maligne et al. [17] also obtained the similar observation.

Figure 12(a) shows fuel film evolution at different injection pressure from 10 to 30 MPa under T_{amb} = 298 K. It is clear to see that the film area increases with higher injection pressure. High injection pressure promotes better atomization, as a result, the splashing droplets may re-impact on the wall with the help of the air entrainment and vortex. Besides, the film on Region I decreases under high injection pressure. It was confirmed by Bai et al. [14, 18] that droplet behaviors change from "stick" to "splash" with lager Weber number. And high injection pressure accelerates velocity thus enlarging Weber number. As a result, more fuel splash off the wall after impingement, leading to less film on Region I. Moreover, the similar conclusion can be drawn from Mie scattering results in **Figures 10** and **11** that both *S* and H_i increase by elevated injection pressure because of droplets splashing. It should be noted that these splashing droplets may re-deposit on the wall to accumulate the film on Region II. Furthermore, film thickness increases with time on both regions as the secondary breakup droplets re-deposit, again.

Figure 12(b) shows fuel film evolution at different injection pressure from 10 to 30 MPa under T_{amb} = 433 K. Same as **Figure 12(a)**, film on Region I decrease with an increase in injection pressure as the splashing behaviors. Compared to T_{amb} = 298 K, fuel film on Region II decreases significantly owing to the fuel evaporation. Moreover, the thickness only becomes larger on Region I. The competition

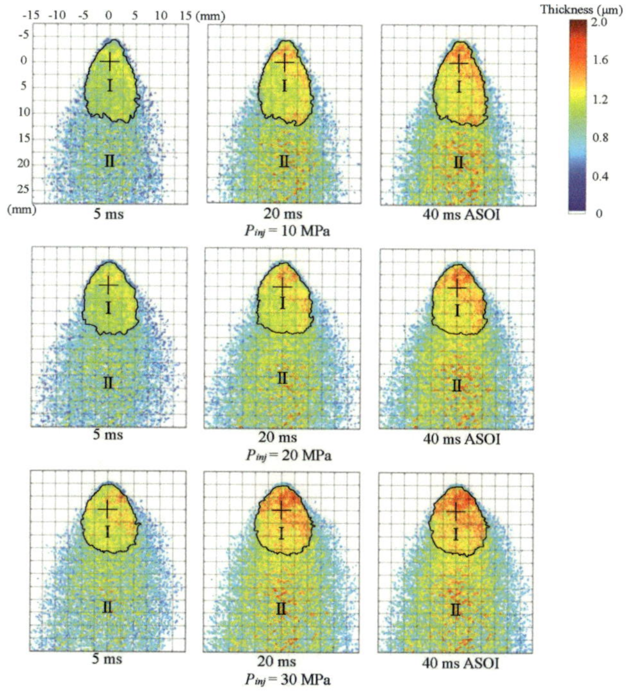

(a)

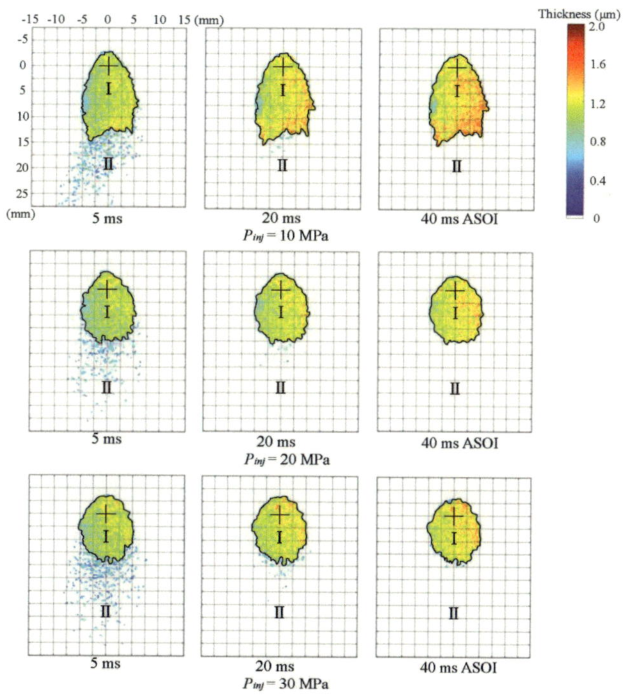

(b)

Figure 12.
Fuel film on the wall. (a) T_{amb} = 298 K. (b) T_{amb} = 433 K.

between droplets depositing and evaporation can be purposed. Maligne et al. [17] observed that film evaporates from thin to thick part due to heat transfer. Under high ambient temperature, the tiny droplets evaporate already before impacting.

Even the secondary droplets re-deposit on Region II to form film, it evaporates very quickly, thus only droplets accumulated on Region I can be observed owing to less heat transfer.

Figure 13 shows the fuel film mass with time after start of injection. The left vertical axis represents film mass, and right one represents the ratio of film mass to injection mass. The results under T_{amb} = 298 K and 433 K are depicted by solid and open data. As we known, two processed that spray development and fuel film formation can be summarized during spray wall interaction [19]. As the strong splashing phenomenon (see from the results of S and H_i above), the maximum only 13% of fuel adheres on the wall under T_{amb} = 298 K. Besides, it provides that film mass increases gradually with time due to the splashing droplets depositing on the wall under T_{amb} = 298 K. However, under T_{amb} = 433 K, the film mass only at P_{inj} = 10 MPa increases slightly with time. The transition changing from "splash" to "stick" tends to be responsible for it. More fuel adheres on Region I after impingement at lower injection pressure, leading to thick film. The splashing droplets re-deposit on Region I, it is difficult to evaporate as thick film existing, thus increasing mass slightly with time. For mass at P_{inj} = 20 and 30 MPa, the initial film mass on Region I is less in contrast to P_{inj} = 10 MPa. Even if the better breakup droplets re-arrive on the wall to form film, it evaporates immediately, which results in a slight decrease of mass. It is interesting to find that high injection pressure increases film mass under T_{amb} = 298 K but decreases film mass under T_{amb} = 433 K. And the opposite phenomena can be explained by different effect of injection pressure on fuel film formation under evaporation and non-evaporation conditions. Under T_{amb} = 298 K, the better atomized droplets accumulate on the wall easily including the first and second adhering on the wall, thus increasing mass at a high injection pressure. However, under T_{amb} = 433 K, the better atomized droplets evaporate easily before impacting on the wall for both incident and re-depositing droplets, leading to few film lefts. Besides, high-speed airflow also facilitates the film evaporation on the wall to decrease it.

Figure 14 describes the film area, and solid and open data are presented to show the results under T_{amb} = 298 K and 433 K, respectively. Similar to mass, film area increases with high injection pressure under T_{amb} = 298 K, but it decreases under T_{amb} = 433 K, which can be attribute to the different effects of injection pressure under ambient temperature conditions. More importantly, as the film evaporates from thin to thick, film area decreases with time, which is different to mass under P_{inj} = 10 MPa.

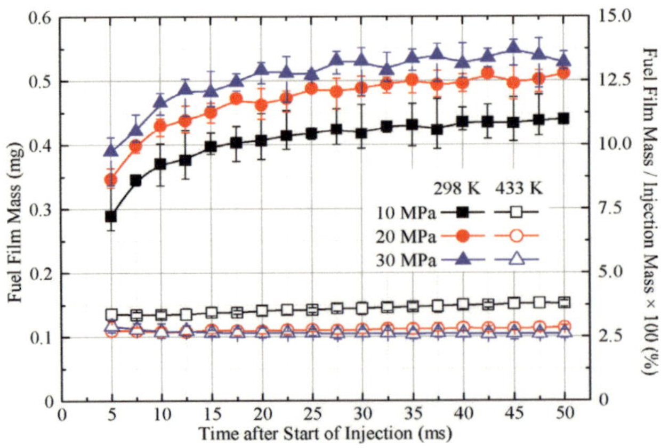

Figure 13.
Fuel film mass under different injection pressures.

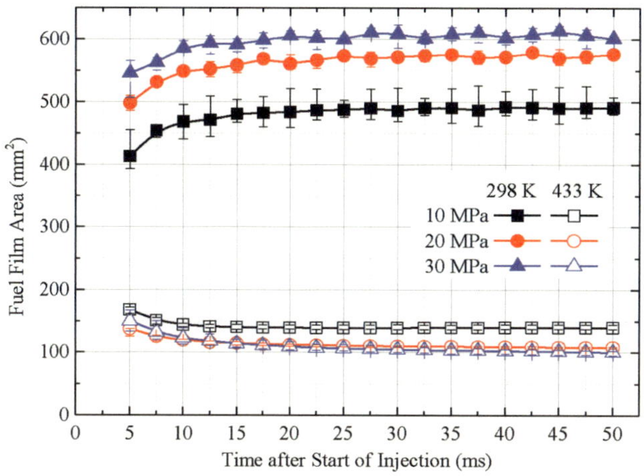

Figure 14.
Fuel film area under different injection pressures.

The film mass distribution at the thickness was defined for deeply study. As illustrated in **Figure 15**, fuel film thickness is presented as the horizontal axis, and film mass is shown by vertical axis. The film mass is calculated by each pixel as shown in the equation below:

$$M(h) = \sum_{i=0}^{\infty} h(i) \tag{1}$$

where the sum of fuel film mass in the thickness fraction between h ± 0.5 Δh and h is defined as $M(h)$, and Δh is 0.05 μm.

Figure 15 shows the film mass along thickness under different conditions at 40 ms ASOI. The peak value decreases dramatically, and curve becomes short owing to the evaporation effect when comparing the results of T_{amb} = 298 K and 433 K. It shows that mass under T_{amb} = 433 K at the thickness from 0.1–0.8 μm decreases to almost 0 in contrast to that under T_{amb} = 298 K, which can be the evidence that high temperature contributes more effort on the thin film evaporation than thick one, thus improving the film uniformity as the narrow shape under T_{amb} = 433 K, which is consistent of the observation in **Figure 12**. Besides, the role of injection pressure can be explained more by checking the smallest peak value at P_{inj} = 10 MPa under T_{amb} = 298 K, but the largest ones under T_{amb} = 433 K.

For better understanding the uniformity of the fuel film, the probability of mass should be used, and it must satisfy the normalization conditions:

$$\sum_{i=0}^{\infty} f_M(h_i) = 1 \tag{2}$$

where the $f_M(h)$ is the probability of $M(h)$.

As illustrated in **Figure 16**, the peak value of probability under T_{amb} = 433 K is much larger than that of T_{amb} = 298 K, indicating more uniform film formed on the wall. No thin film at the periphery should be the reason for it due to evaporation. It is interesting to find that under T_{amb} = 433 K, although peak value of mass occurs at P_{inj} = 10 MPa, but the probability at P_{inj} = 30 MPa becomes the largest one, suggesting high injection promotes film uniformity under evaporation condition. However, under non evaporation condition, less influence of injection pressure can be detected.

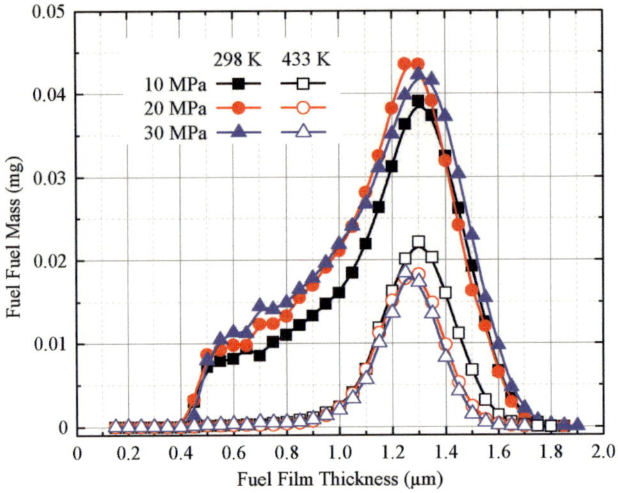

Figure 15.
Fuel film mass distribution under different injection pressures.

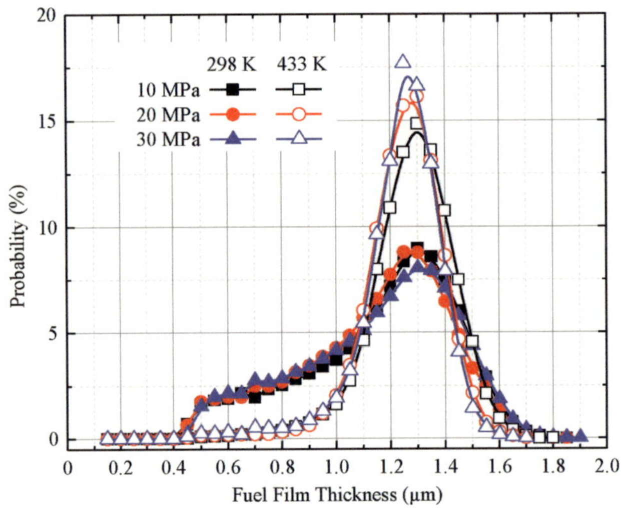

Figure 16.
Probability of film thickness under different injection pressures.

3.3 Effect of ambient pressure

Figure 17 shows the S under different ambient pressures, with results of T_{amb} = 298 and 433 K depicted by solid and open symbols. It suggests that the increased ambient pressure decreases S, and the raised density of the air results in stronger air-fuel entrainment and less spray momentum should be the reason for that under both non-evaporation and evaporation conditions. H_i under different conditions are compared in **Figure 18**. At the initial stage from 0.2 to 1.0 ms ASOI, H_i is larger under low ambient pressure. H_i generates easily as less resistance from ambient pressure should be the reason for it. While, H_i becomes larger under high ambient pressure with spray development. The raised ambient gas density from high ambient pressure results in strong air-fuel entrainment. Therefore, more air-fuel mixture can be expected after impingement, thus increasing H_i under high ambient pressure. In contrast to that, after EOI, the spray propagates along the wall

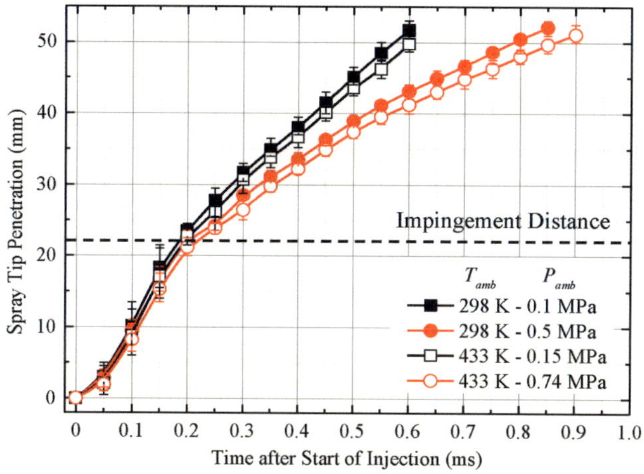

Figure 17.
Spray tip penetration under different ambient pressures.

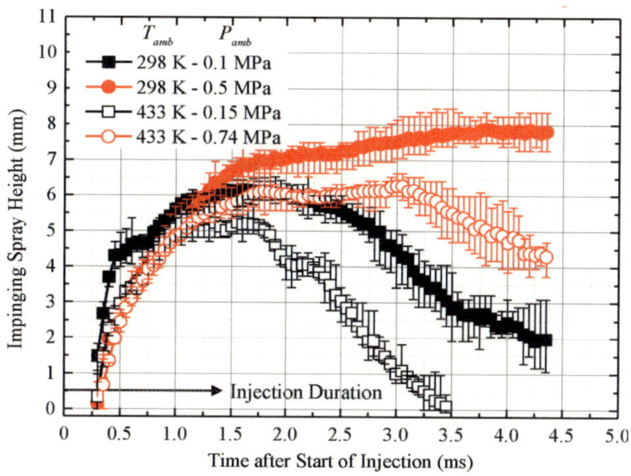

Figure 18.
Impinging spray height under different ambient pressures.

quickly with less air drag force under low ambient pressure, leading to no fuel supply. Therefore, H_i decrease sharply after EOI. Besides, the high ambient temperate promotes this behavior. However, owing to the slow air-fuel movement under high ambient pressure, H_i can keeps constant and even increases to some extent after EOI, suggesting high ambient pressure enlarges H_i. One thing should be noted that although the high ambient pressure suppresses H_i, the behaviors of slow movement accumulating H_i cannot be hindered.

In order to discuss the effect of ambient pressure on the fuel film evolution, different timing at 5, 20, and 40 ms ASOI are selected to show at ambient pressures under T_{amb} = 298 K and 433 K, as shown **Figure 19**. The cross symbol is used to present the impingement point, and color bar is applied to show the thickness from 0 to 2.0 μm. It has already illustrated in **Figure 9** that two regions on the fuel film on the wall could be separated with different formation mechanism as Regions I and II. And the ambient pressure has more influences on the film formation on Region II.

Figure 19(a) illustrates that fuel film on both Regions I and II increases with time elapse, and the re-depositing droplets accumulating on the wall should be the reason for it. Moreover, it has to be noted that the film area increases with high ambient pressure. Three main reasons can be summarized. Firstly, elevated ambient pressure promotes better atomization, and these atomized droplets re-arrive on the wall easily by vortex to increase film. Secondly, high ambient pressure decelerates fuel spray before impingement, leading to decreasing Weber number. Therefore, the droplet behaviors may change from "splash" to "spread" or "stick" with small Weber number, causing adhered film increase. Thirdly, under high ambient pressure, the splashing droplets could collide and coalesce with others easily, and finally re-deposit on the wall to increase the fuel film.

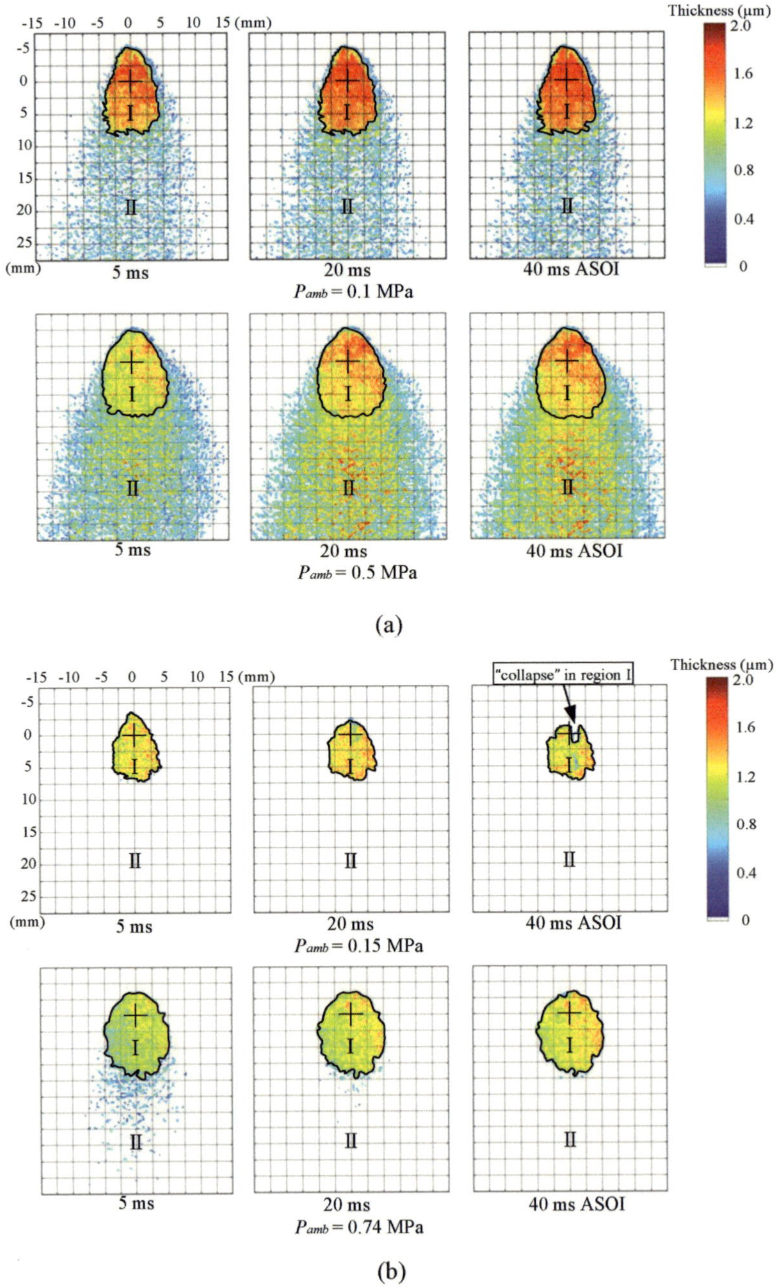

Figure 19.
Evolution of fuel film on the wall. (a) T_{amb} = 298 K. (b) T_{amb} = 433 K.

Figure 19(b) illustrates the fuel film evolution on the wall at T_{amb} = 433 K. It is clear that owing to evaporation, fuel film area decreases obviously with time. Furthermore, more fuel locates on Region I other than Region II, and the less heat transfer at thick film should be the reason for it. Under T_{amb} = 433 K, some of the splashing droplets evaporate before re-impacting on the wall. And the survived re-depositing droplets on Region II evaporate immediately as strong hear transfer, leading to more fuel film accumulating on Region II at last. While, the survived ones impact on Region I then form film on it successfully. It can be seen that fuel film exist on Region II at 5 ms ASOI under P_{amb} = 0.74 MPa, and the large droplets re-depositing and adhering on the wall before evaporation should be the main reason for it. Besides, the fuel film area under P_{amb} = 0.74 MPa is larger than that under P_{amb} = 0.15 MPa. In addition to the reasons discussed above, high ambient pressure hinders the film evaporation to a certain extent could be another reason. One more interesting thing is that under P_{amb} = 0.15 MPa, fuel film evaporates from periphery until 40 ms ASOI, and the "collapse" appearing in the film affects the film uniformity, which will be discussed more in the following part.

Figure 20 shows the fuel film mass with time after start of injection. The left vertical axis represents film mass, and right one represents the ratio of film mass to injection mass. The results under T_{amb} = 298 K and 433 K are depicted by solid and open data. It depicts that film mass increases with time gradually under T_{amb} = 298 K. And the re-depositing droplets after splashing should be responsible for it. Noticed that the increased amplitude of film mass under P_{amb} = 0.5 MPa is much larger when compared to of P_{amb} = 0.1 MPa. On one hand, high ambient pressure enlarges and decelerate droplets, leading to more droplets tend to re-deposit on the wall. On the other hand, the behavior transition from "splash" to "stich" increases film on the wall. But looking at the mass under T_{amb} = 433 K, it decreases with time slightly due to evaporation. Even if some re-deposited droplets adhere on the wall, the evaporation effect cannot be ignored. More importantly, mass under P_{amb} = 0.74 MPa is larger than that under P_{amb} = 0.15 MPa. As explained in the experimental condition, T_{sat} (398 K) < T_{amb} under P_{amb} = 0.15 MPa, indicating a strong film evaporation. While, T_{sat} (472 K) is larger than T_{amb} under P_{amb} = 0.74 MPa, and a slow evaporation can be expected. Recently, Schulz et al. [20] found that the time for film evaporation duration increases at elevated ambient pressure, implying that high

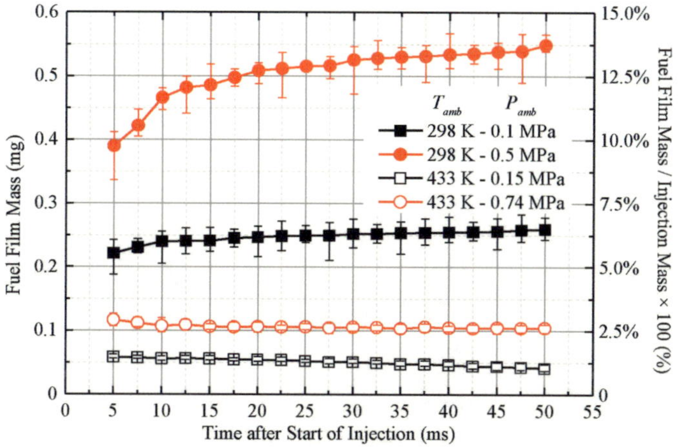

Figure 20.
Fuel film mass under different ambient pressures.

ambient pressure has a significant effect on inhibiting evaporation of fuel film. This founding is confirmed by Kim et al. [21] from their single droplet experiment and Tao et al. [22] from the computational fluid dynamics (CFD)results. Furthermore, the results that film mass under P_{amb} = 0.74 MPa is larger than that under P_{amb} = 0.15 MPa can also demonstrate that high ambient pressure inhibits the fuel evaporation.

Figure 21 describes the film area, and solid and open data are presented to show the results under T_{amb} = 298 K and 433 K, respectively. As the re-depositing droplets impact on the same location, leading to large increase in mass but little increase in area. Therefore, fuel film under T_{amb} = 298 K increases with time slightly. Same as film mass, the area increases with high ambient pressure. Under non-evaporation condition, the large size, low velocity, "stick" behavior and easy re-deposition should be attribute to it. Under evaporation condition, in addition to the above factors, the inhibition of film evaporation under high ambient pressure should be another reason for it.

Figure 22 shows the film mass along thickness under different conditions at 40 ms ASOI. And the results under T_{amb} = 298 K and 433 K are shown by solid and

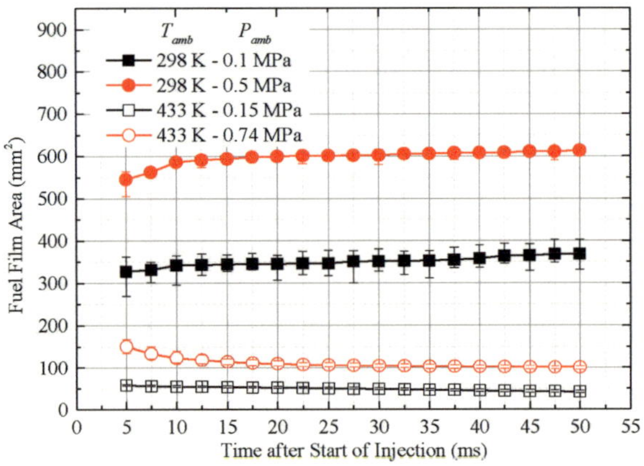

Figure 21.
Fuel film area under different ambient pressures.

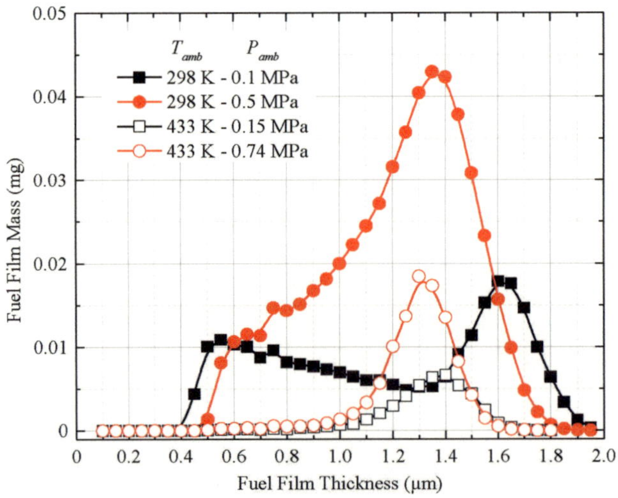

Figure 22.
Fuel film mass distribution under different ambient pressures.

open symbols, respectively. It reveals that the peak value increases significantly with high ambient pressure. It should be clarified that under T_{amb} = 298 K, tow peak values at 0.5 and 1.6 μm can be observed for P_{amb} = 0.1 MPa condition, indicating film accumulated on Region II and the central of region I. However, by increasing the ambient pressure to 0.5 MPa, it is interesting to find these two peaks merge into one, indicating more uniform film form under high ambient pressure.

Figure 23 shows the probability of film mass along thickness under different conditions at 40 ms ASOI, and the results under T_{amb} = 298 K and 433 K are depicted by solid and open symbols, respectively. The probability increases sharply under evaporation condition, indicating high temperature facilitates uniform film on the wall. And the evaporation occurs from thin to thick should be the main reason. Special attention should be taken that not only ambient temperature, but also ambient pressure can improve the fuel film uniformity. Elevating the ambient pressure at a certain temperature can increase the density of the ambient gas, leading to a strong interaction between air gas and fuel. As a result, fuel spray angle increases, and droplets kinetic energy decreases. Eventually, the droplets attach o more evenly on the wall, thereby forming a uniform film on the wall. Because of this reason, under T_{amb} = 298 K, when ambient pressure increased from 0.1 to 0.5 MPa, these two peaks merge into one, indicating large improvement for film uniformity. Similarly, under T_{amb} = 433 K, as less film located on Region II, the two peaks disappear. By increasing the ambient pressure from 0.15 to 0.74 MPa, peak value increases, suggesting uniformity improved, although not so obvious as that under non-evaporation condition.

To sum up, **Figure 24** displays the relationship between the saturation vapor pressure curve and the tested conditions. It shows that above the curve toluene is under the liquid phase but transits to vapor phase below the curve. By increasing the ambient pressure from 0.1 to 0.5 MPa under T_{amb} = 298 K, the ambient density increases. Then, the stronger entrainment and air resistance can be expected, leading to low velocity and large size of the droplets, and behavior changing from "splash" to "stick", which are the reasons for fuel film increase. In addition, increased the ambient pressure from 0.15 to 0.74 MPa under T_{amb} = 433 K, because of the transient from vapor to liquid phase, the decrease in evaporation rate as one more reason increases fuel film on the wall.

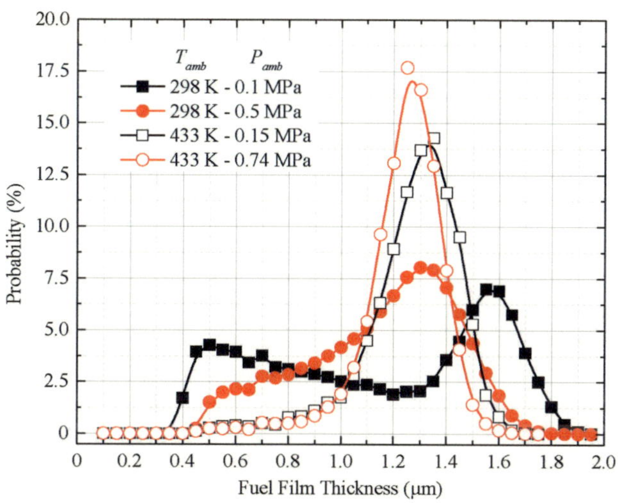

Figure 23.
Probability of film thickness under different ambient pressures.

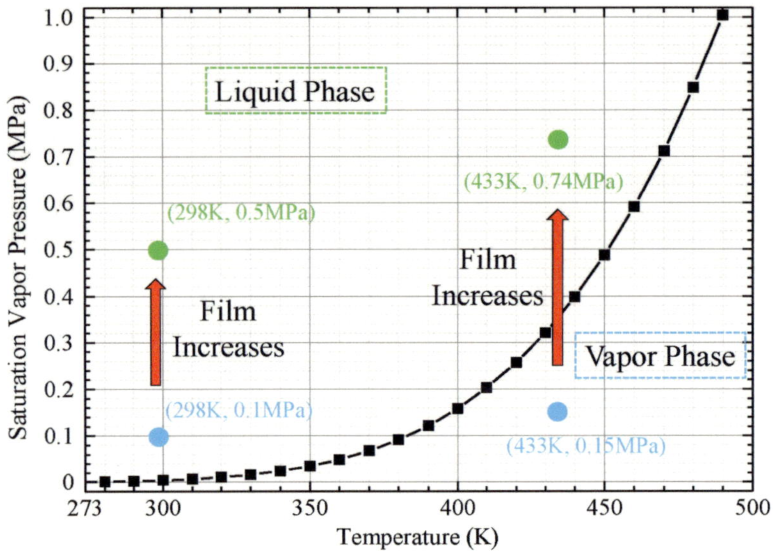

Figure 24.
Saturation vapor curve of tested fuel.

3.4 Fuel film evaporation characteristics

Figure 25(a) shows the fuel film evolution at P_{amb} = 0.15 MPa. Different timings at 50, 70, 100, and 150 ms ASOI are selected to depict and the color bar presents thickness from 0 to 2.0 μm with cross symbol being as impingement point (*o*). Owing to the fuel evaporation, the film area decreases with time. Furthermore, the film evaporates from the cross symbol to downstream, making the fuel film "moves downward". The film thickness increases from up to downstream due to the interaction of fuel spray and wall, leading to the fuel film evaporates quickly at the impingement region. As a result, it appears the film moving downward. Besides, the increased injection pressure decreases film area significantly. Three reasons can be concluded. Firstly, the high injection promotes better atomization, resulting in more tiny droplets evaporate before impacting on the wall. Secondly, even if some droplets survive to impact on the wall, with larger Weber number, the trend of "splash" after impingement can be expected. Thirdly, even though fuel finally sticks on the wall, the air flow with high velocity improves the evaporation rate, leading to less film left. Noteworthy is that the film evaporates from the periphery to the central of film firstly as Maligne and Bruneaux [17] reported, then evaporates from the central causing a "hollow" in the film, which will be analyzed in the following part. It is interesting that under the elevated injection pressure, more film can be observed near the cross symbol, especially at 150 ms ASOI. The increased injection pressure enlarges Weber number, leading to more fuel "splash" instead of "spread". Finally, less fuel accumulates at the downstream.

Figure 25(b) shows the fuel film evolution at P_{amb} = 0.74 MPa. More fuel film can be seen left on the wall when ambient pressure increased to 0.74 MPa. The same as **Figure 25(a)**, it is evident that fuel film area decreases sharply against high injection pressure. In addition, the increased ambient pressure also increases film area. The droplets are decelerated by stronger air force of high ambient pressure. Therefore, more droplets tend to "stick" on the wall, resulting in more film formed. Besides, even some droplets splash off the wall, the splashing droplets tend to collide and then coalesce with others. Finally, these larger droplets redeposit on the wall more easily, thus accumulating more film. Moreover, the evaporation rate decreases under high ambient pressure, leading to more film left. And more discussions about the evaporation rate

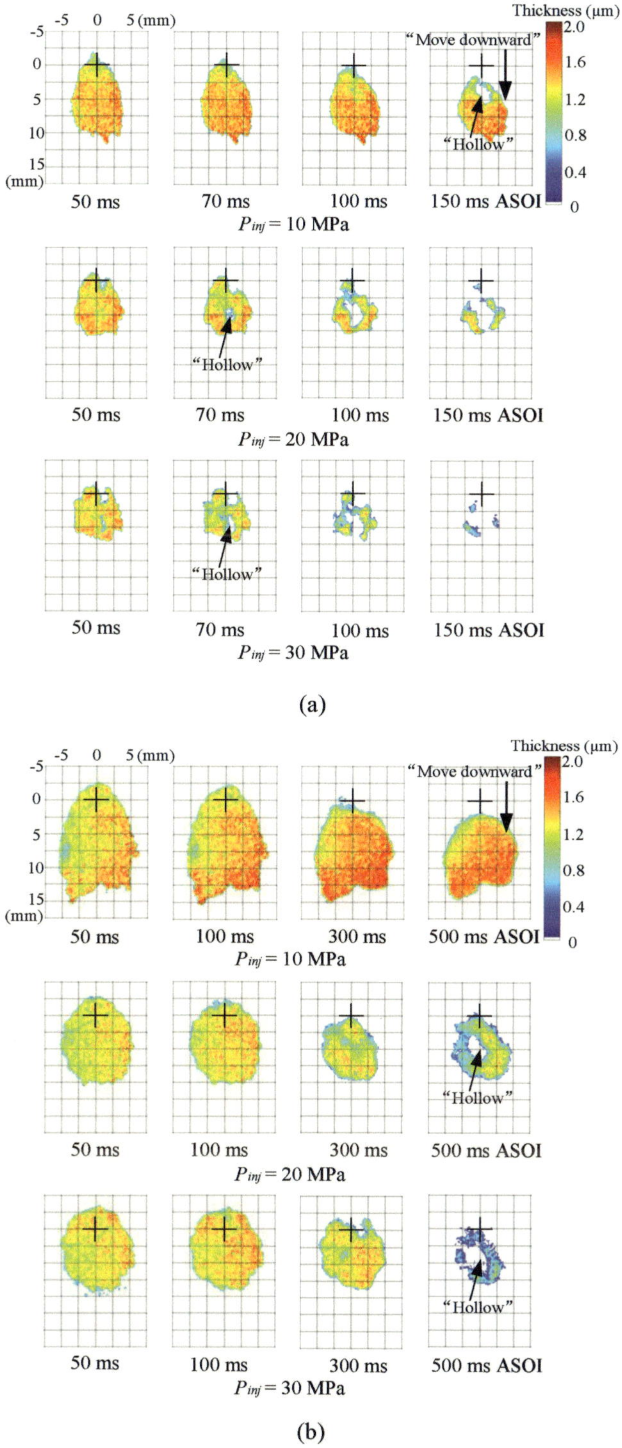

Figure 25.
Evolution of fuel film. (a) P_{amb} = 0.15 MPa. (b) P_{amb} = 0.74 MPa.

will be discussed in the next part. Besides, similar to **Figure 25(a)**, more fuel can be seen near the cross symbol with an increase in injection pressure at 500 ms ASOI.

Figure 26 shows fuel film mass under different conditions with left vertical axis being mass and right one being the ratio of film mass to injection mass. Results

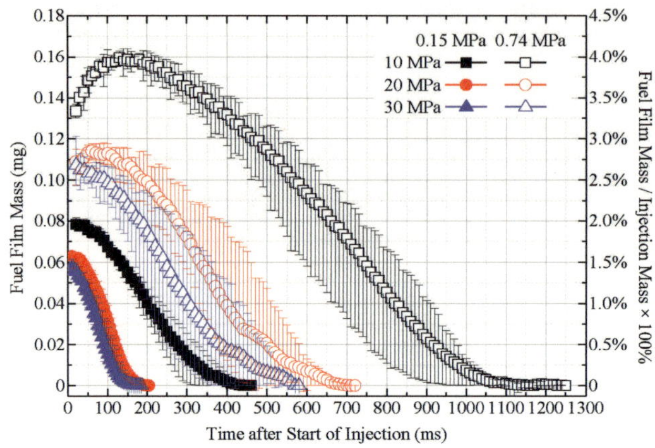

Figure 26.
Evaporation characteristics of fuel film mass.

under P_{amb} = 0.15 and 0.74 MPa are shown by filled and open data. It is evident that the mass increases with an increase in injection pressure due to the better atomization and fast evaporating rate. While, the mass increases with an increase in ambient pressure. More fuel sticking on the wall and low evaporating rate should be responsible for it. It should be noted that the film mass firstly increases and then decreases under only one case of P_{inj} = 10 MPa, P_{amb} = 0.74 MPa. Although the injection duration is only 2.9 ms, mass increases until 200 ms ASOI. It should be attributed to the lowest velocity of liquid droplets and largest size of them. As described above, under high ambient but low injection pressures condition, the spray is decelerated causing the impacting droplets arrive at the wall relatively late, which is the main reason for mass increasing gradually. Moreover, droplets tend to coalesce into larger more after impingement making it difficult to evaporate. In addition, droplets re-deposit on the wall easily under this case. It is interesting to see that the maximum ratio is approximately 4%, while the minimum one is 1.5%, suggesting most fuel evaporates or splashes off the wall.

Figure 27 shows fuel film area under different conditions. Results under P_{amb} = 0.15 and 0.74 MPa are shown by filled and open data. Similar to mass, the elevated injection pressure enlarges film area decreases but ambient pressure increases it with the same reason as mass. More importantly, different to mass, no obvious increase in area can be seen under P_{inj} = 10 MPa, P_{amb} = 0.74 MPa, implying that the impacting droplets arrive at the same location, leading to only mass increase. Besides, the lifetime of fuel film can be obtained from **Figures 26** and **27**, which will be discussed more in **Figure 28**.

Figure 29 provides the details about the "hollow" development in fuel film under P_{inj} = 10 MPa, P_{amb} = 0.15 MPa. It reveals that film evaporates at periphery from 60 to 120 ms ASOI, causing it "moving downward" as analyzed in **Figure 25**. Generally, evaporation occurs during the film formation, also leading to the film becomes thin with time for the whole region. At around 140 ms, it shows "hollow" occurs. With "hollow" development, film becomes less uniform, leading to quick evaporation at "hollow" until all evaporated.

A conceptual model for fuel film evaporation is proposed here to describe its mechanism and evaporation dynamic, as shown in **Figure 30**. Three processed can be summarized "Evaporate from periphery - Hollow occur - Evaporate from hollow". Because the fuel film at the periphery is relatively thin, strong evaporation can be expected owing to heat transfer in contrast to central part. With the consideration of the flat wall used with roughness, the film is non-uniform initially. Then,

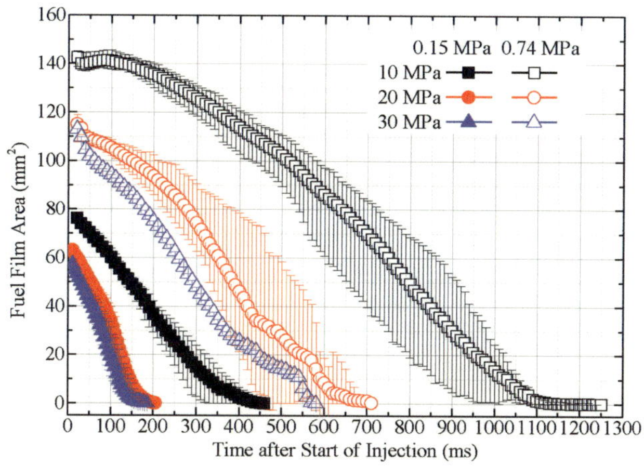

Figure 27.
Evaporation characteristics of fuel film area.

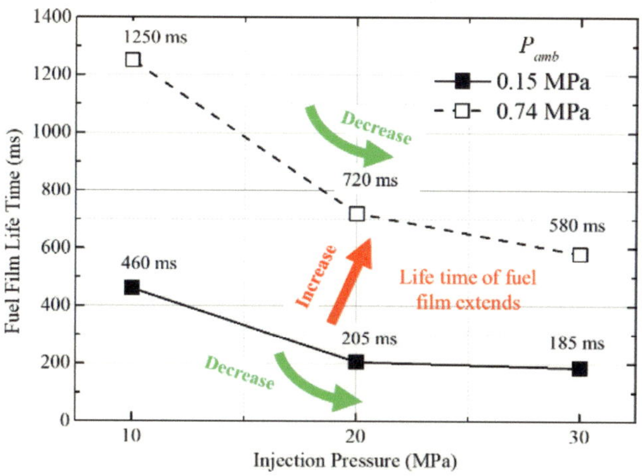

Figure 28.
Fuel film lifetime.

"hollow" appears in the film with evaporation, leading to film "breakup". With the "hollow" development, film evaporation is facilitated, and film becomes less uniform until all evaporates finally.

Figure 28 illustrates the fuel film lifetime under different conditions. Noting that the lifetime is calculated from the start of injection until all film evaporates, as shown in **Figures 26** and 27. One more thing should be noted that although the lifetime is quite longer than one cycle of the real working condition in gasoline engine, lifetime is an important factor to evaluate film evaporation characteristics, especially for validating the simulation result. The observation shows that injection pressure decreases film lifetime owing to the enhanced evaporation rate and better atomization. Moreover, the decreased lifetime from 10 to 20 MPa is more obvious than that from 20 to 30 MPa, suggesting limitation may exist for the high injection pressure shortening lifetime. But ambient pressure increases the lifetime mainly due to the liquid/vapor transition by elevating ambient pressure from 0.15 to 0.74 MPa. The phase transition can decrease the evaporation rate sharply, leading to longer lifetime of the film. Besides, this may be another reason for the mass increases with high ambient pressure. But no influence can be found from the injection pressure on phase changing.

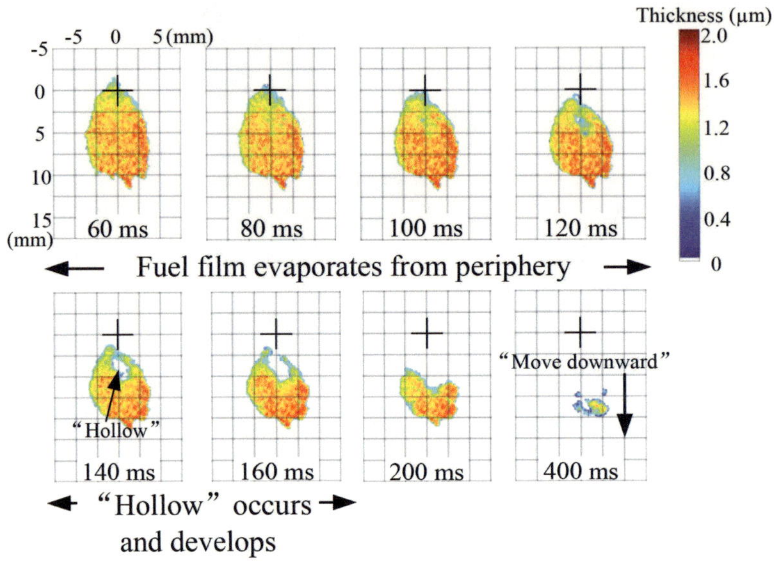

Figure 29.
"Hollow" development.

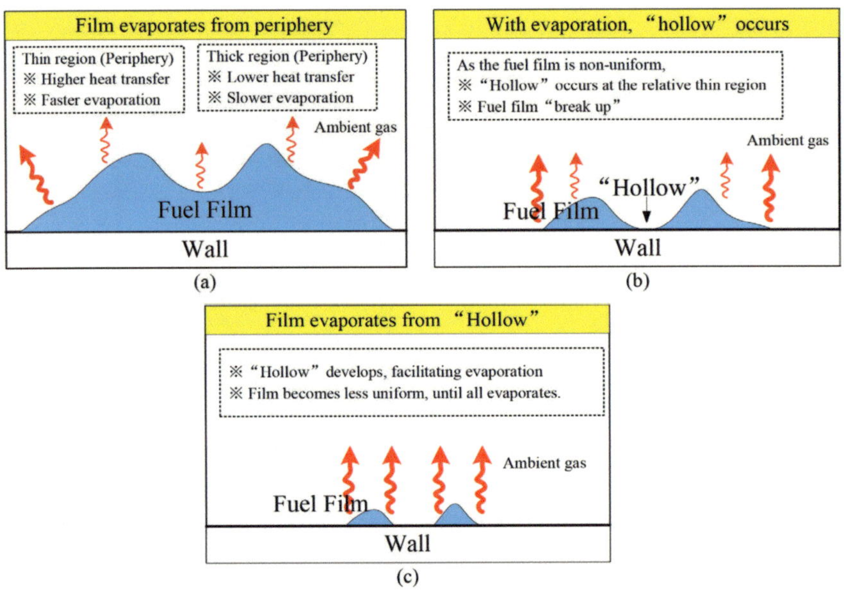

Figure 30.
Schematic conceptual model. (a) Film evaporates from periphery. (b) With evaporation, "hollow"occurs. (c) Film evaporates from "hollow".

4. Conclusions

This chapter discussed about the fuel spray and impingement under non-evaporation and evaporation conditions. The impinging spray development and fuel film formation as well as evaporation characteristics were checked and compared, the major conclusions are summarized as follows:

1. Different impingement regions were characterized through the formation of fuel film, known as the primary impingement region (called Region I) and the

secondary impingement region (called Region II). Furthermore, high ambient temperature exerts more influence on the fuel film formation for the Region II.

2. During the injection, a high injection pressure accelerates the fuel spray and favors the atomization of fuel, thus enlarging S and H_i of liquid phase under both T_{amb} = 298 K and T_{amb} = 433 K. Under T_{amb} = 298 K, injection pressure increases the fuel film mass and area. However, under T_{amb} = 433 K, injection pressure decreases the them, and the better atomization and easy evaporation of the fuel are the main reasons for it.

3. The increased ambient pressure increases the density of the gas, resulting in strong air-fuel entertainment and less momentum, thus shortening S and increasing H_i. Moreover, higher ambient pressure increases the fuel film mass and area under both T_{amb} = 298 K and T_{amb} = 433 K.

4. The fuel film firstly evaporates from periphery, leading to fuel film "moving downward". Then, "hollow" occurs and develops, resulting in less uniformity of fuel film. Finally, the fuel film evaporates from the "hollow" until all evaporated. The fuel film lifetime is shortened by increasing injection pressure, but it extends with an increase in ambient pressure.

Acknowledgements

The author would like to acknowledge National Natural Science Foundation of China [51909037] and State Key Laboratory of Clean Energy Utilization [ZJU-CEU 2019005].

Nomenclature

CFD	Computational fluid dynamics
DISI	Direct injection spark ignition
RIM	Refractive index matching
PFI	Port fuel injection
UHC	Unburned hydrocarbon
PM	Particulate matter
PN	Particle number
EV	Electric vehicle
T_{sat}	Saturated temperatures
fps	Frames per second
S	Spray tip penetration
H_i	Impinging spray height
L_w	Distance to wall
L_r	Radial distance
ASOI	After start of injection
EOI	End of injection
T_{amb}	Ambient temperature
P_{inj}	Injection pressure
P_{amb}	Ambient pressure

Author details

Hongliang Luo
Hiroshima University, Higashi-Hiroshima, Japan Foshan University, Foshan, P.R. China

*Address all correspondence to: luo@hiroshima-u.ac.jp

References

[1] Maly RR. State of the art and future needs in SI engine combustion. Symposium (International) on Combustion 1994; 25(1): 111-124. https://doi.org/10.1016/S0082-0784(06)80635-1

[2] Drake MC and Haworth DC. Advanced gasoline engine development using optical diagnostics and numerical modeling. Proceedings of the Combustion Institute 2007; 31(1): 99-124. https://doi.org/10.1016/j.proci.2006.08.120

[3] Kalantari D, Tropea C. Spray impact onto flat and rigid walls: Empirical characterization and modelling. International Journal of Multiphase Flow 2007; 33(5): 525-544. https://doi.org/10.1016/j.ijmultiphaseflow.2006.09.008

[4] Tanaka D, Uchida R, Noda T, Kolbeck A, Henkel S, Hardalupas Y, Taylor A, Aradi, A. Effects of fuel properties associated with in-cylinder behavior on particulate number from a direct injection gasoline engine. SAE Technical Paper, 2017-01-1002; 2017. https://doi.org/10.4271/2017-01-1002

[5] Zhao F, Lai MC, Harrington DL. Automotive spark-ignited direct-injection gasoline engines. Progress in energy and combustion science 1999; 25(5): 437-562. https://doi.org/10.1016/S0360-1285(99)00004-0

[6] Moreira ALN, Moita AS, & Panao MR. Advances and challenges in explaining fuel spray impingement: How much of single droplet impact research is useful?. Progress in energy and combustion science 2010; 36(5): 554-580. https://doi.org/10.1016/j.pecs.2010.01.002

[7] Park SW and Lee CS. Macroscopic and microscopic characteristics of a fuel spray impinged on the wall. Experiments in fluids 2004; 37 (5): 745-762. https://doi.org/10.1007/s00348-004-0866-3

[8] Andreassi L, Ubertini S, & Allocca L. Experimental and numerical analysis of high pressure diesel spray–wall interaction. International journal of multiphase flow 2007; 33(7): 742-765. https://doi.org/10.1016/j.ijmultiphaseflow.2007.01.003

[9] Guo M, Shimasaki N, Nishida K, Ogata Y, Wada Y. Experimental study on fuel spray characteristics under atmospheric and pressurized cross-flow conditions. Fuel 2016; 184:846-855. https://doi.org/10.1016/j.fuel.2016.07.083

[10] Luo H, Uchitomi S, Nishida K, Ogata Y, Zhang W, Fujikawa T. Experimental Investigation on Fuel Film Formation of Spray Impingement on Flat Walls with Different Surface Roughness. Atomization and Sprays, 2017; 27(7): 611-628. DOI: 10.1615/AtomizSpr.2017019706

[11] Luo H, Nishida K, Ogata Y. Evaporation characteristics of fuel adhesion on the wall after spray impingement under different conditions through RIM measurement system. Fuel, 2019;258: 116163. https://doi.org/10.1016/j.fuel.2019.116163

[12] Luo H, Nishida K, Uchitomi S, Ogata Y, Zhang W, Fujikawa T. Effect of Spray Impinging Distance on Piston Top Fuel Adhesion in Direct Injection Gasoline Engines. International Journal of Engine Research, 2020; 21(5):742-754. https://doi.org/10.1177/1468087418774175

[13] Luo H, Nishida K, Uchitomi S, Ogata Y, Zhang W, Fujikawa T. Effect of temperature on fuel adhesion under spray-wall impingement condition.

Fuel 2018, 234:56-65. https://doi.org/10.1016/j.fuel.2018.07.021

[14] Bai C, Gosman AD. Development of methodology for spray impingement simulation. J Engines 1995,104(3):550-68. https://www.jstor.org/stable/44633238

[15] Mundo CHR, Sommerfeld M, Tropea C. Droplet-wall collisions: experimental studies of the formation and breakup process. Int J Multiph Flow 1995;21(2):151-73.

[16] Ding CP, Sjöberg M, Vuilleumier D, Reuss DL, He X, Böhm B. Fuel film thickness measurements using refractive index matching in a stratified-charge SI engine operated on E30 and alkylate fuels. Exp Fluids 2018;59(3):59. https://doi.org/10.1007/s00348-018-2512-5

[17] Maligne D, Bruneaux G. Time-resolved fuel film thickness measurement for direct injection SI engines using refractive index matching. SAE Tech Paper 2011. 2011-01-1215. https://doi.org/10.4271/2011-01-1215

[18] Bai C, Rusche H, Gosman AD. Modeling of gasoline spray impingement. Atom Sprays 2002;12 (1-3):1-27. DOI: 10.1615/AtomizSpr.v12.i123.10

[19] Baumgarten C. Mixture Formation in Internal Combustion Engines. Germany: Springer Verlag; 2006.

[20] Schulz F, Beyrau F. Systematic Investigation of Fuel Film Evaporation. SAE Technical Paper 2018-01-0310; 2018. https://doi.org/10.4271/2018-01-0310

[21] Kim H, Sung N. The effect of ambient pressure on the evaporation of a single droplet and a spray. Combust Flame 2003;135(3):261-70. https://doi.org/10.1016/S0010-2180(03)00165-2

[22] Tao M, Ge H, VanDerWege B, Zhao P. Fuel wall film effects on premixed flame propagation, quenching and emission. Int J Engine Res 2018. (Online First). https://doi.org/10.1177/1468087418799565

Chapter 3

Vehicles Power Consumption: Case Study of Dar Rapid Transit Agency (DART) in Tanzania

Kenedy Aliila Greyson

Abstract

Energy consumption and its environmental impact are now among the most challenging problems in most developing cities. The common sources of energy used as the fuel in transportation sector include gasoline, diesel, natural gas, propane, biofuels, electricity, coal, and hydrogen. However, in Tanzania, diesel and gasoline are still the dominant source of energy used by public and private vehicles. We have experienced significant efforts of converting conventional vehicles (gasoline engines) to operate on Compressed Natural Gas (CNG) or on hybrid system (gasoline and natural gas) as an alternative source of energy in Tanzania. The CNG is considered as cleaner combustion energy used as a vehicular fuel alternative to gasoline or diesel. In this chapter, the amount of energy consumption from the fuel combustion, the impact of environmental health (toxicity gas emission), the cost of fuel used by the transit buses in terms of fuel energy consumption, and driving profile are discussed. The scope of this work is based on the total energy contained in the fuel only. The ability of the engine to transform the available energy from the fuel into useful work power (efficiency) is left to the designers and manufacturers.

Keywords: CNG, consumption, cost, diesel, emission, energy, gasoline

1. Introduction

The Tanzania strategic plans to increase economy are focused on industrialization agenda. The implementation of this agenda will increase the energy demand, hence the heavy fuel consumptions in industries and transportations sector. Energy consumption and its environmental impact are now among the most challenging problems in the most cities. Dar es Salaam is one of those fastest growing cities in Tanzania with the population over 5 million and expected to be a mega city with more than 10 million by 2030. Traffic congestion is one of the most prevalent transport challenges in Dar es Salaam. Due to the traffic congestion there is an increase of fuel consumption and likely potential harmful emissions. Moreover, the high fuel consumption (the imported commodity) significantly affects the national economy. The strategies are to reduce the harmful emissions by using fuel with less emission such as Compressed Natural Gas (CNG) and to develop electric powertrain systems. The chapter concludes with a brief look at fuel consumption and carbon dioxide (CO_2) emissions in public transport vehicles and other modes of transport.

The Government of Tanzania through Dar es Salaam Rapid Transit (DART) agency operates a Bus Rapid Transit (BRT) system in Dar es Salaam. BRT system aimed to be a cost effective sustainable transportation system for the city of Dar es Salaam to ensure fast and orderly flow of traffic on urban streets and roads. It is reported by DART agency that, in phase one project each trunk bus and feeder bus daily operate nearly 297 km and 245 km, respectively. However, BRT's bus fare is still higher than other operators (daladala) due to the high operation costs. DART projects are implemented in phases. According the DART Agency [1], the phase one of DART corridor mainly involved 20.9 km of roads: Kimara to Kivukoni (15.8 km), Magomeni to Morocco (3.4 km) and Fire to Kariakoo (1.7 km). Other facilities include 27 bus stations and 5 terminals. The phase two is along Kilwa Road from city centre to Mbagala Rangi Tatu; other parts of the corridor start from South Kawawa road at Morogoro Road junction to Mgulani JKT round with 20.3 km. The implementation of the phase three DART project goes through the Nyerere highways from Gongolamboto to the City Center and some area of the Uhuru Road from Tazara to Kariakoo to Gerezani, which covers a total of 23.6 km.

Motor vehicles have been powered by gasoline, diesel, steam, gas turbine, and electric. [2]. The common Internal Combustion (IC) engines are gasoline spark-ignition and diesel compression-ignition engines involving the combustion of a fuel inside a chamber that results in the expansion of the air/fuel mixture to produce mechanical work. In Tanzania, gasoline is the most common ground-transportation fuel, mostly used by private vehicles, followed by diesel used by most public vehicles and heavy duty trucks. As it is for the most public buses, BRT buses also use diesel as a fossil fuel. Diesel fuel in compression ignition engines produces a high level of toxicity in emission gases which leads to a health and environmental hazard [3]. The emitted gases associated with diesel fuel include nitrous oxides (NO_x), carbon monoxide (CO), and carbon dioxide (CO_2).

Tanzania has huge reserves of Natural Gas that city buses, especially the BRT systems, can use so as to minimize the harmful emissions. The development of gas supply infrastructures in Dar es Salam, Mtwara, and Coast Region are the examples of the ongoing initiatives. Recently, the natural gas, CNG in particular, has been considered to be a potential replacement and alternative to diesel and petrol fuels in vehicles due to its lower hazardous emission of gases in the environment. Applications of natural gas have been given much attention among stakeholders and researchers due to the remarkable attributes towards greener transportation. Studies on the use of natural gas in trains, buses, trucks, motorcycles, scooters and bicycles have been presented in various research reports. Since Tanzania has huge reserves of natural gas, this market growth increases with the users in various sectors including transportation.

Natural gas is used in transportation sector (public and private vehicles), industries, and domestic. If the use of natural gas resource is adopted effectively, the life of the people will improve both economically and environmentally. Therefore, a well-designed plan for the use of the natural gas in transmission and distribution is important. The objective is to optimize the utilization of gas resource as the alternative fuel energy in various sectors and the environmental safety. The strategy is to emphasize on the supply of CNG for transport sector use, and Piped Natural Gas (PNG) for domestic, commercial and industrial sectors in Tanzania. The CNG is mostly used in a public transit transportation fuel in other countries. CNG is a processed fluid gas into a high-pressure natural gas compression in a tube [4]. Compared to gasoline and diesel, vehicles powered by CNG emit less carbon monoxide, nitrogen oxides (NO_x), and particulate matter [5]. Apart from the government initiatives of using CNG in transportation, there are strategic plans to supply

CNG to the industries outside the downstream pipeline, and to other customers for heating and cooking purposes.

Since both the gas and petrol engines run according to the Otto principle with a spark plug ignition, there are several centers in Tanzania converting petrol engines to work in hybrid systems. Dar es Salaam Institute of Technology (DIT) and University of Dar es Salaam (UDSM) are among these centers. It is therefore very important that the plans for the natural gases infrastructure expansions get real so that suppliers and operators can invest in proper natural gases products to be used effectively in transportation, industry and domestic sectors. Currently, at CNG stations, gas flow to the vehicle is measured by mass for sale with dispensers designed with compensation for temperature variations to ensure accurate quantities are delivered to the tank [6].

Since, transit buses are among the cost effective forms of mass transit; most cities consider transit buses as the backbone of the transport system. However, the environmental health impact and running cost must be closed managed. In this chapter we investigate the current situation of BRT system buses using diesel, CNG and gasoline engines. The fuel consumption which is mentioned as the factor that affects the running costs, the cost of the fuel and the amount of fuel per distance, driving profile, and the emitted gases (mainly carbon dioxide) are discussed.

2. Propulsion energy

In this section, we briefly consider the main energy sources; gasoline, diesel and CNG as the input power to the IC engine vehicles. The fuel consumption, and overall efficiency, on the baseline scenario are discussed. Although urban driving is characterized by lower driving speeds and presumably more frequent stops which lead to higher consumption values [7, 8], this study is characterized by the special lanes (that is, not shared by other vehicles) for DART's buses. In this section, the energy from the CNG, gasoline and diesel fuels are analyzed where the difference in (potential) energy of reactants (exothermic reaction) and products are presented.

2.1 Natural gas

The heat is given off when a specified amount (say, one mole) of a methane (reactant) burns in oxygen gas (heat of combustion) during the reaction. The general bond dissociation energy equation (enthalpy change) for the reaction at a specified temperature through the reaction is converted into a motive force for the vehicle is in shown in Eq. (1) [9]. The heating of the fossil fuel results in CO_2 emission.

$$C_xH_y + \left(x + \frac{y}{4}\right)O_2 \rightarrow xCO_2 + \frac{y}{2}H_2O + \text{Energy} \tag{1}$$

where C_xH_y is the generic chemical formula for the fossil fuel, and O_2, the oxygen in the input, and the output of the reaction are the emitted carbon dioxide (CO_2), water (H_2O) and the released energy.

Methane (natural gas), which is a fossil fuel burns in oxygen to release energy as shown in Eq. (2). From **Table 1**, the combustion of methane gas will release 802.3 KJ/mol which is equivalent to calorific values of 50.1 KJ/g. The energy content of different fuels is made up of different chemical compounds, that is, bond breaking energies and bond forming energies reaction.

$$CH_4{}^{(g)} + 2O_2{}^{(g)} \rightarrow CO_2{}^{(g)} + 2H_2O^{(g)} + \text{Energy} \tag{2}$$

Bond Type	Energy (kJ/mole)
C–H	414
C–C	347
C=C	615
O–H	463
C–O	360
C=O	728
H–H	436
O=O	498

Table 1.
Average bond energies of common bonds.

2.2 Gasoline

Energy from the gasoline (petrol) released during the reaction is as shown in Eq. (1). From the general reaction, the octane (reactant) burns in oxygen gas to release energy from the reaction. The bond dissociation energy (enthalpy change) for the reaction at a specified temperature through the gasoline chemical reaction is shown in Eq. (3).

$$2C_8H_{18}{}^{(l)} + 25O_2{}^{(g)} \rightarrow 16CO_2{}^{(g)} + 18H_2O^{(g)} + \text{Energy} \tag{3}$$

The calculation reveals that the calorific value of the gasoline (petrol) is approximately 42 KJ/g.

2.3 Diesel

Energy from diesel is released during the reaction shown in Eq. (1) where the dodecane reactant, in the case of diesel, burns in oxygen gas to release energy from the reaction. The bond breaking energies and bond forming energies reaction is shown in Eq. (4).

$$4C_{12}H_{23}{}^{(l)} + 71O_2{}^{(g)} \rightarrow 12CO_2{}^{(g)} + 46H_2O^{(g)} + \text{Energy} \tag{4}$$

The calculation reveals that the calorific value of the diesel is approximately 43 KJ/g.

3. Longitudinal dynamics of the vehicle

Vehicle dynamics depend on tire and road contact forces and torques, mass of the vehicle, road profile (road grade angles), ambient conditions, and driving profile. These parameters contribute into the total fuel consumption. The total power consumption is defines by the vehicle dynamics and distance travelled by the vehicle. The IC engine vehicles transform the chemical energy (from gasoline/ petrol, diesel, or CNG) into useful work (power consumed). The energy released by the gasoline, diesel, and CNG driven vehicles is discussed here. In fact, there are various factors affecting energy consumption when using IC engines. The ratio of power transmitted in a rotating shaft, referred to as brake power, and engine input

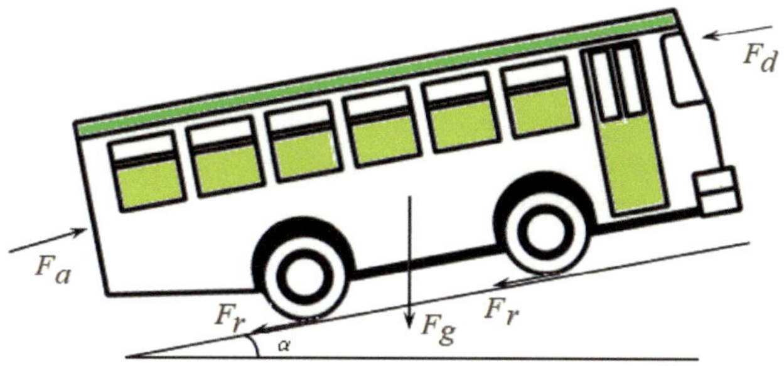

Figure 1.
Forces acting on a moving BRT bus.

power provides the overall efficiency of the IC engine. It should be noted that, modern engines (petrol and diesel) have better efficiency. **Figure 1** shows the forces acting on the moving bus. Therefore, the discussion assumes the better performance of engines.

3.1 Internal combustion engine power

It has been mentioned that BRT buses utilizes diesel as a fossil fuel. In analyzing the power consumed by the BRT bus, we need to understand the components of forces and resistance (opposing forces) that act on it. These forces and resistance on the bus are; rolling resistance due to tire and road interaction, aerodynamic drag resistance, grade resistance depends on motion towards the up-hill or the down-hill, and accelerating force due to the motion of the vehicle mass as shown in Eq. (5)-(8) [10–13].

$$F_r = c_r mg \cos\alpha \tag{5}$$

$$F_d = \frac{1}{2}\rho c_d A v^2 \tag{6}$$

$$F_g = mg \sin\alpha \tag{7}$$

$$F_a = ma \tag{8}$$

where,
F_r is the rolling resistance,
F_d is the aerodynamic drag resistance,
F_g is the grade resistance,
F_a is accelerating force,
c_r is the rolling resistance coefficient,
m is the vehicle mass (N),
α is the road grade angle (radian),
c_d is the wind resistance coefficient $(N/(m^2/s)^2)$,
A is the vehicle frontal area (m^2),
v is the vehicle speed (m/s),
a is the vehicle acceleration (m/s^2),
ρ density of air, and.
g is the gravitational constant (nominally 9.81 m/s^2).

By definition, the rolling resistance is the combination of all frictional load forces of the tire on the road surface and the friction within the BRT bus. Aerodynamic drag is the resistance of air to the movement of the vehicle. The gradability of the vehicle load power can increase or decrease depending on whether the car is ascending or descending an incline. Therefore, the vehicle power needed are typically based on vehicle acceleration requirements, usually specified as the time to accelerate and depends on the maximum available torque and maximum available power of the propulsion system. Then, the prime mover force which is the total tractive force, F_T *for* the vehicle is given in Eq. (9).

$$F_T = F_r + F_d + F_a + F_g \tag{9}$$

The tractive power and total tractive energy are given in (10) and (11), respectively.

$$P = F_T v \tag{10}$$

$$E = Pt = P\frac{s}{v} \tag{11}$$

where,
t is the time spent,
s is the distance travelled,
P is the tractive power, and.
E is the tractive energy.

4. BRT bus power consumption

The BRT system scenario and power consumption are presented in this section. The type of buses and routes are depicted in **Figures 2** and **3**, respectively. Our discussion of the vehicle's engine performance will consider the DART phase one network (route), from Kivukoni terminal to Kimara terminal with 21 stations and 3 terminals. The DART's buses in this route start from Kivukoni station to Kimara in Dar es Salaam as a one-segment of the route. For analysis purpose, diesel fuel, gasoline fuel, and CNG fuel are considered in the discussion section. Technically, the speed of the vehicle, v (km/h) is obtained to estimate the speed of the engine, ω (rpm) and torque, τ (Nm).

Table 2 presents the route codes and destinations of the BRT buses in DART project phase one. The route code 001, from Kivukoni to Kimara, is used in this discussion. Other routes can be analyzed in the approach. Distances between

Figure 2.
BRT bus at the station.

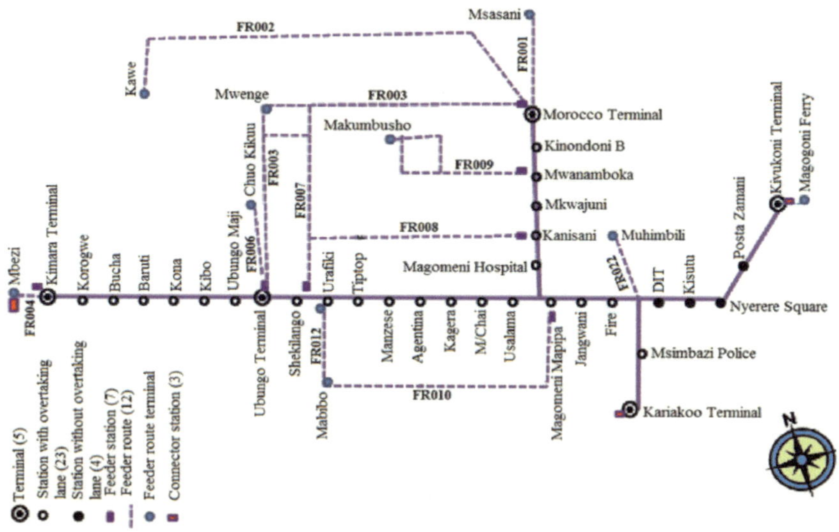

Figure 3.
DART phase one network Source: DART agency [1].

SN	Route Code	Source	Destination
1	001	Kimara	Kivukoni
2	002	Ubungo	Kivukoni
3	003	Morocco	Kivukoni
4	004	Kimara	Gerezani
5	005	Ubungo	Gerezani
6	006	Morocco	Gerezani
7	007	Kimara	Morocco
8	008	Ubungo	Morocco
9	080	Muhimbili	Gerezani
10	090	Kimara	Mbezi

Table 2.
DART project phase one route codes and destinations.

stations and terminals are presented in **Table 3**. The total distance covered by the route code 001 is 15.6 km. The estimated energy consumption is calculated by using Eq. (11) based on the road profiles (grade angles at each sample) and driving profiles (speed, stops, accelerating, and braking). Then, the estimated energy consumption is converted into kWh to be compared with the available energy in the fossil fuels [14]. This consumption depends on the motion of the vehicle and how it behaves in motion. Due to the fact that, these dynamic parameters are complex and varies from different inputs, the discussion is focused only on the energy released from the fuel as the input of the engines.

4.1 Driving profile measurements

Measurement of the driving profile can be done in different ways. The rotation vector sensor and the gravity sensor are the most frequently used sensors for motion detection and monitoring. Speed, orientation and position of the DART's

SN	Name	Category	Distance (km)
1	Kivukoni	Terminal	0.0
2	Posta Zamani	Station	1.14
3	Nyerere Square	Station	1.52
4	Kisutu	Station	2.25
5	DIT	Station	2.62
6	Fire	Station	3.18
7	Jangwani	Station	3.85
8	Magomeni Mapipa	Station	4.88
9	Usalama	Station	5.6
10	Mwembe chai	Station	6.12
11	Kagera	Station	6.6
12	Argentina	Station	7.25
13	Bakrhesa	Station	7.77
14	Manzese	Station	8.4
15	Urafiki	Station	9.0
26	Shekilango	Station	9.8
17	Ubungo	Terminal	10.5
18	Ubungo Maji	Station	11.2
19	Kibo	Station	12.2
20	Kona	Station	12.8
21	Baruti	Station	13.38
22	Bucha	Station	13.86
23	Korogwe	Station	14.5
24	Kimara	Terminal	15.8

Table 3.
DART Route 001 distances between stations/terminals.

bus were recorded by the use of the Global Positioning System (GPS) and Inertial Navigation System (INS) where integration of Inertia Measurement Unit (IMU) and GPS (GPS/IMU) are used [15].

There are several techniques to obtain the speed of the moving vehicles. For example, algorithm to estimate vehicle speed from accelerometer data generated by an onboard smart-phone [16] has been proposed in various literatures. However, according to authors in [16], speed estimation by the integration of the mobile-phone accelerometer data will not yield accurate results, since the accelerometer data in the direction of motion is not pure acceleration, but involves white noise, phone sensor bias, vibration, gravity component, and other effects. Similarly, the use of GPS to measure vehicle speed, on the other hand, depends on the number of visible GPS satellites at the recording time. Other factors include model of the mobile phone, whether condition, radio noise, concurrent usage, location and type of terrain, mobile network triangulation, Wi-Fi network location, and magnetic fields. [17] depending on the algorithm used. The speed of the vehicle can be estimated after analyzing the acceleration characteristic values in advancing direction and vertical direction as explained in [18].

4.2 Road profile measurements

The road profile and vehicle's speed (discussed earlier) are measured to determine power consumed by the vehicle. The quality of the road profile defines comfort ride, quality of load and passenger transportation as well as external excitation of the ground vehicle [19]. This subsection presents the road condition monitoring where rotation vector sensor and the gravity sensor are used.

The road grade angle (even vehicle speed) can be obtained by this method. The algorithm using KALMAN filters combines the use of accelerometer sensor and gyroscope (also known as gyro) sensor readings to obtain the speed and road profile. The readings in all orientations (x, y, z) are modeled as shown in Eqs. (12)-(14) where x_{xi} is the reading along x dimension, x_{yi} is the reading along y dimension, and x_{zi} is the reading along z dimension of the respective sensor (accelerometer and gyroscope). The t is the time of reading at k sample.

$$g_x^{(k+1)} = g_x^{(k)} + \nu_x^{(k)} t \tag{12}$$

$$g_y^{(k+1)} = g_y^{(k)} + v_y^{(k)} t \tag{13}$$

$$g_{y1}^{(k+1)} = g_{y1}^{(k)} + v_{y2}^{(k)} t \tag{14}$$

where,

v_x, v_y and v_z are sensor readings in x, y, and z orientations at t interval of time, and g_x, g_y and g_z are the road grade or speed in x, y, and z orientations at k sample time and updated at $k + 1$ sample.

Finally, the readings at each sample are utilized to compute the road profile.

5. Results discussions

The engine torque is created on the crankshaft by the cylinder pressure pushing on the piston during the power stroke. Therefore, the maximum torque depends on the pressure pushed during the power stroke. However, in practice, the torque versus speed characteristic of an IC engines are not as linear as electric machines.

The typical gasoline engine operates at no more than 10:1 compression ratio while the typical diesel engine may operate with a compression ratio as high as 25:1. It is obvious that, the higher the compression ratio, the better the overall engine efficiency. However, the scope of this work is based on the total energy contained in the fuel only. The great deal of the energy produced by a combustion engine is wasted. In this work, it is assumed that the speed of the vehicle along the route is the same regardless of the engine used.

The driving profile (time/speed) of the BRT bus for route 001 obtained is shown in **Figure 4**. The distance of 15.8 km was travelled in about 37 minutes. The average speed of this BRT bus is 26 kilometer per hour (kph). Despite the 23 expected stops at the stations and terminals, there are other stops at the road junctions, and zebra crossings. Moreover, the drivers try to observe the speed limit of 50 kph and 60 kph along the route 001 road. The power consumption is proportional to the modes of the vehicle: traction (when the prime mover force, $F_t > 0$), braking ($F_t < 0$), and coasting ($F_t = 0$). **Figure 5** shows the average power consumption for the BRT bus on route 001. The energy used to brake or slow down a vehicle in a conventional vehicle is dissipated as heat in the braking system and lost to the vehicle. An electric vehicle can be designed to regenerate the energy and store it on the vehicle for auxiliary usage.

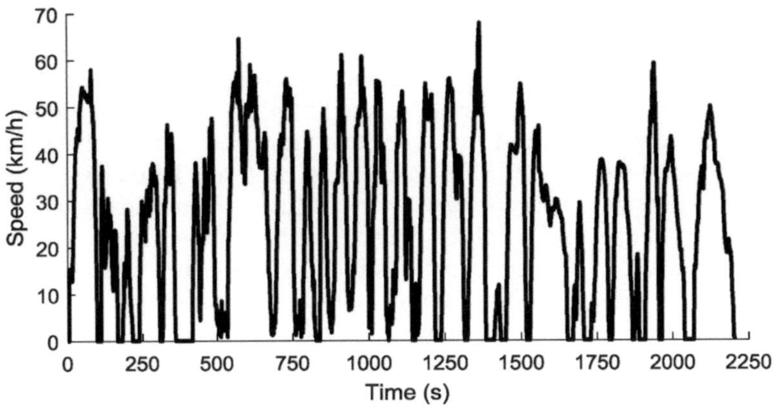

Figure 4.
Driving profile of the BRT bus on route 001.

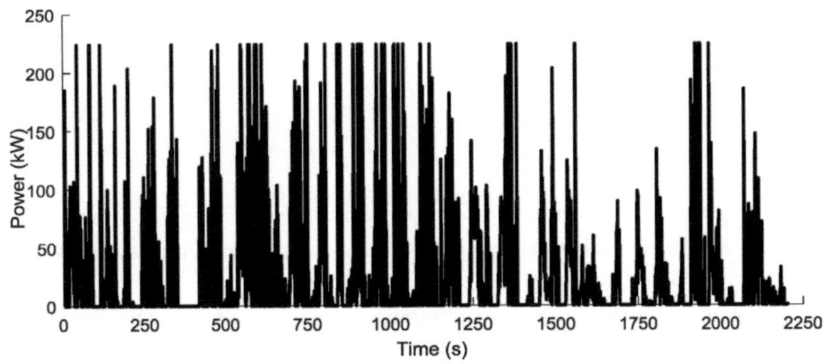

Figure 5.
Average power consumption for the BRT bus on route 001.

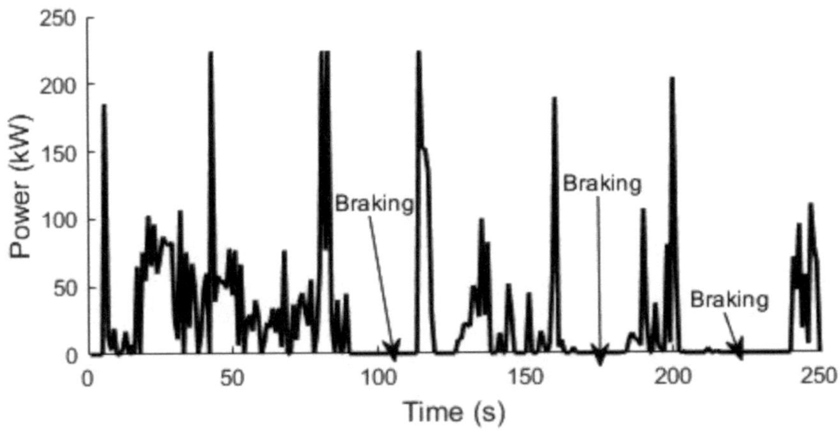

Figure 6.
Section of power consumption (0–250 second).

Figure 6 depicts the section of power consumption (0–250 seconds), showing the stopping (or braking) periods. The more power is consumed when the bus is in accelerating mode. As discussed earlier, the power consumption (and emission) depends on the vehicle design parameters, auxiliary devices, driving profiles, road profiles and conditions. **Figure 7** shows the amount of energy consumption for the BRT bus on route 001 obtained by using Eq. (11).

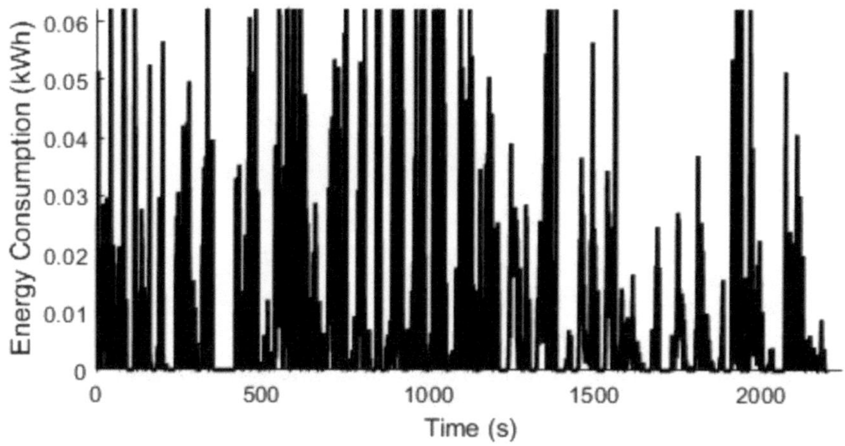

Figure 7.
Energy consumption for the BRT bus on route 001.

Parameter	Unit	Value
Wind resistance coefficient	$N/(m^2/s)^2$	0.599
Rolling Resistance Coefficient, C_r	—	0.01
Vehicle mass, m	kg	11750
Passenger mass	kg	5590
Vehicle total mass	kg	17340
Frontal area, A	m2	6.93
Acceleration due to gravity, g	m/s^2	9.8
Final-drive ratio, r_f	—	3.37
4-speed transmission ratios, r_t	—	[2.393 1.450 1.000 0.677]
Wheel radius, r_r	m	0.381
Engine efficiency	—	0.92
Tank capacity is about	liter	475
Maximum engine torque	—	1200 Nm
Maximum engine power	—	284 kW

Table 4.
Transit bus parameters.

The total energy consumed when the given vehicle (parameters in **Table 4**) on route 001has been computed to be approximately 19.952 kWh. In fossil fuel environment, this energy can be obtained from gasoline, diesel, or CNG. The calorific values of these fuels (Energy/mass) can be in KJ/g or kWh/Kg. In converting calorific value, energy/mass, in KJ/g of the fuel into calorific value in kWh/ Kg, we divide the energy/mass (KJ/g) value by 3.6. Gasoline and diesel densities are presented in **Table 5** [20]. Using information from **Table 5**, other conversions (kWh/Kg and kWh/L) are presented in **Table 6**.

In Tanzania mainland, the Energy and Water Utilities Regulatory Authority (EWURA) is responsible to publish cap prices for petroleum products. The wholesale and retail prices vary from time to time depending on price of the imported petroleum product imported. For example, According to the released wholesale and retail prices released by EWURA on Wednesday, 3rd February 2021, the retail

Fuel	Kg/liter
Gasoline	0.737
Diesel	0.850

Table 5.
Specific gravity of motor fuel (Kg/L).

Fuel	KJ/g	kWh/Kg	kWh/L
Gasoline	42	11.67	8.6
Diesel	43	11.94	10.15
CNG	50.1	13.92	—

Table 6.
Other units for energy mass of the fuel.

Fuel	Required energy (kWh) route 001		Fuel Cost*
	Kg	Liters	
Gasoline	—	2.32	4,377.84
Diesel	—	1.9	3,475.10

Gasoline (petrol): 1887.00 TZS/Liter; Diesel 1829.00 TZS/Liter; CNG: 1550.00 TZS/kg.

Table 7.
Energy consumed and cost of fuel used in route 001.

Fuel	CO_2 Emission (Kg)
Gasoline	5.336
Diesel	5.073
CNG	3.257*

Emission when high calorific gas is used.

Table 8.
CO_2 emission in one segment of Kivukoni-Kimara route.

prices were as indicated in the underneath of **Table 7**. From Eq. (11) the average required energy for the BRT bus on route 001 approximately 19.952 kWh. This energy can be delivered by the gasoline, diesel or CNG as shown in **Table 7**. Therefore, with other factors assumed the same and perfect, gasoline IC engine bus would consume 2.32 liters of gasoline to cover the route while diesel IC engine would consume 1.9 liter of diesel in the same route. CNG engine bus, on the other hand, would 1.22 Kg of CNG gas. The total fuel cost required is shown in the extreme column of **Table 7**.

The emission of carbon dioxide (CO_2) for an internal combustion engine to move a vehicle down the road depends on the type of fuel used. While 1 L of gasoline produces approximately 2.3 Kg of CO_2, burning 1 L of diesel produces approximately 2.67 Kg of CO_2. For the case of CNG, CO_2 emission depends on the calorific gas (i.e. low calorific gas and high calorific gas). This emission ranges from 2.25 Kg of CO_2 (low calorific gas) to 2.67 Kg of CO_2 (high calorific gas) when burning 1 Kg of CNG. Hence, the total CO_2 emission in route 001 is shown in **Table 8**. The CNG emission is still low compared to gasoline and diesel.

6. Conclusion

The concerns of the global warming and environmental pollution have led to more severe regulations on CO_2 and other pollutant emissions. Therefore, converting conventional vehicles to operate on natural gas is a good option for BRT system. The operating cost is low due the low cost of CNG and the CO_2 gas emission is low compared to other fuel types (diesel and gasoline). Since CNG is available locally, (not imported), the national agenda of industrialization is supported by the use of CNG as an alternative source of energy.

Author details

Kenedy Aliila Greyson
Dar es Salaam Institute of Technology (DIT), Dar es Salaam, Tanzania

*Address all correspondence to: kenedyaliila@yahoo.com

References

[1] United Republic of Tanzania President's Office, Regional Administration and Local Government, Dar Rapid Transit (DART) agency, Available from: https://www.dart.go.tz [Accessed: 2020-07-12].

[2] Committee on the Assessment of Technologies for Improving Light-Duty Vehicle Fuel Economy, National Research Council. Assessment of fuel economy technologies for light-duty vehicles. Washington, DC: The National Academies Press, 2011. 14 p. DOI: 10.17226/12924.

[3] Kojima M. Breathing Clean: Considering the switch to natural gas buses. World Bank Technical Paper No. 516, 2001; p.1-50, DOI: 10.1596/0-8213-5040-4.

[4] Isworo P. Compressed natural gas technology for alternative fuel power plants," in Proc. 2nd International Conference on Energy, Environmental and Information System (ICENIS), August 14-15, 2018; Semarang, Indonesia. p. 1-4.

[5] Speight J. G. Natural Gas A Basic Handbook, 2nd ed. Elsevier Inc.; 2019. p. 20-21. DOI: 10.1016/C2015-0-02190-6.

[6] Kagiri C, Zhang L, Xia X. Optimization of a compressed natural gas station operation to minimize energy cost. In Proceedings of 9th International Conference on Applied Energy (ICAE2017), 21-24 august 2017; Cardiff, the United Kingdom: Elsevier, 2017. p. 2003-2008.

[7] Wang H, Zhang X, Ouyang M. Energy consumption of electric vehicles based on real-world driving patterns: A case study of Beijing. Applied Energy, 2015; 157(1):710–719. DOI: 10.1016/j.apenergy.2015.05.057.

[8] Braun A, Rid W. Energy consumption of an electric and an internal combustion passenger car. A comparative case study from real world data on the Erfurt circuit in Germany. In: Proceedings of 20th EURO Working Group on Transportation Meeting, (EWGT 2017); 4-6 September 2017; Budapest, Hungary. Transportation Research Procedia 2017; p. 468–475.

[9] John H, Abas Goodarzi G. Electric Powertrain: Energy Systems, Power Electronics and Drives for Hybrid, Electric and Fuel Cell Vehicles. Wiley; 2019. DOI: 10.1002/9781119063681.

[10] Sousa D, Costa J, Dente J. Electric bicycle using batteries and supercapacitors. In: Proceedings of the European Conference on Power Electronics and Applications; 2-5 September 2007; Aalborg, Denmark, 2007. p. 774 – 781.

[11] Wai K, Rong Y, Morris S. Simulation of a distance estimator for a battery electric vehicle. Alexandria Engineering Journal; 2015; 54 (3):359-371. DOI: 10.1016/j.aej.2015.04.008.

[12] Łebkowski A. Studies of energy consumption by a city bus powered by a hybrid energy storage system in variable road conditions. In Energies. 2019; 12: 951. DOI: 10.3390/en12050951.

[13] Zongxuan S, Guoming Z. Design and Control of Automotive Propulsion Systems. Boca Raton, Fla:CRC Press, 2015. DOI: 10.1201/b17947.

[14] Wai, K, Rong, Y, Morris S. Simulation of a distance estimator for a battery electric vehicle. Alexandria Engineering Journal. 2015; 54 (3): 359-371. DOI: 10.1016/j.aej.2015.04.008.

[15] Rogers M, Applied Mathematics in Integrated Navigation Systems. 3rd ed.

American Institute of Aeronautics and Astronautics; 2003.p.246. DOI: 10.2514/4.861598.

[16] UstunI, Cetin M. Speed estimation using smart-phone accelerometer data. Transportation Research Record. 2019; 2673(3): 65–73. DOI: 10.1177/0361198119836977.

[17] Sakperea W, Adeyeye-Oshinb M, Mlitwa W. A state-of-the-art survey of indoor positioning and navigationsystems and technologies. South African Computer Journal. 2017; 29(3):145–197. DOI: DOI: 10.18489/sacj.v29i3.452.

[18] Zong, X,Wen X. A new approach to estimate real-time traveling speed with accelerometer. International Journal of Distributed Sensor Networks.205; 11 (10) DOI:10.1155/2015/928168.

[19] Yunusov A, Eshkabilov S, Riskaliev D, Abdukarimov N. Estimation and evaluation of road roughness via different tools and methods. In: Proceedings of XI International Scientific Conference; 26-28 June 2019; Poland. p. 770-784.

[20] Simetric. Specific Gravity of Liquids. 2020. Available from: http://www.simetric.co.uk/si_liquids.htm [Accessed: 2020-10-12]

Section 2

Application of Biodiesel

Chapter 4

Shock Tube Combustion Analysis

Claudio Marcio Santana and Jose Eduardo Mautone Barros

Abstract

The shock tube is a metal tube that the gas at low pressure and high pressure are separated by a diaphragm. When the diaphragm (make of material copper and aluminum) breaks on predetermined conditions (high pressure in this case) produces shock waves that move from the high-pressure chamber (known the compression chamber or Driver section) for low pressure chamber (known the expansion chamber or Driven section). The objective of this work is the correlate the ignition delay times of convectional Diesel and Biodiesel from soybean oil measured in a shock tube. The results were correlated with the cetane number of respective fuels and compared with the ignition delay times of Diesel and Biodiesel with cetane numbers of known. The ignition delay time of biodiesel from soybean oil was approximately three times greater than the ignition delay time of convectional Diesel. The contribution of this work is that it shows why pure biodiesel should not be used as substitutes for Diesel compression ignition engines without any major changes in the engines.

Keywords: diesel ignition delay time, biodiesel ignition delay time, shock tube, diesel cetane number, biodiesel from soybean oil cetane number

1. Introduction

Shock tube is an equipment used to study gas flow in different areas of engineering and operating conditions, such as: shock wave movement, aerodynamic flows under different temperature and pressure conditions, gas compressibility and fuel combustion. The equipment is constructed by a metal tube separated by a diaphragm, which divides the equipment into two sections. The high-pressure section is called the driver section while the low-pressure section is called the driven section. The diaphragm separating the two sections is designed to withstand a certain pressure, when that pressure is reached the diaphragm breaks and a compression wave is formed and moves towards the driven section. Instantly an expansion wave is formed and propagates towards the driver section. This movement of the gas mass inside the shock tube causes an increase in pressure and temperature in the driven section and a reduction in pressure and temperature in the driver section, [1, 2]. The **Figure 1** shows the driver and driven sections of shock tube before the rupture of the diaphragm.

After the diaphragm rupture is formed inside the shock tube, the contact surface, a region that did not feel the passage of the shock wave, a second region that felt the effects of the passage of the shock wave, a third region that did not feel the passage of the expansion wave and a fourth region that felt the effects of the expansion wave passage, [2, 4]. The **Figure 2** shows the contact surfaces and four regions formed after the diaphragm rupture.

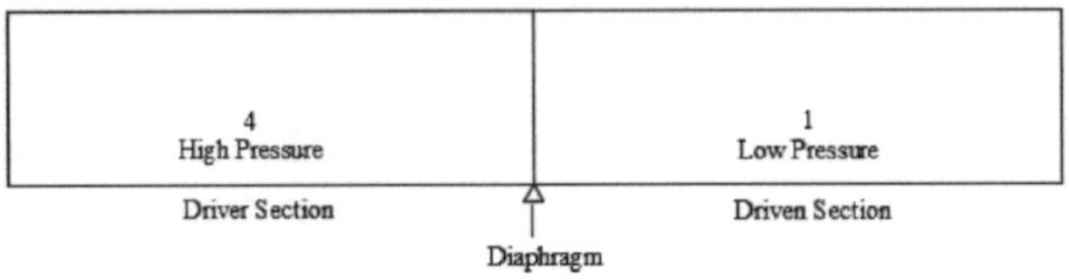

Figure 1.
Driver and driven sections of the shock tube before rupture of the diaphragm (adapted from [3]).

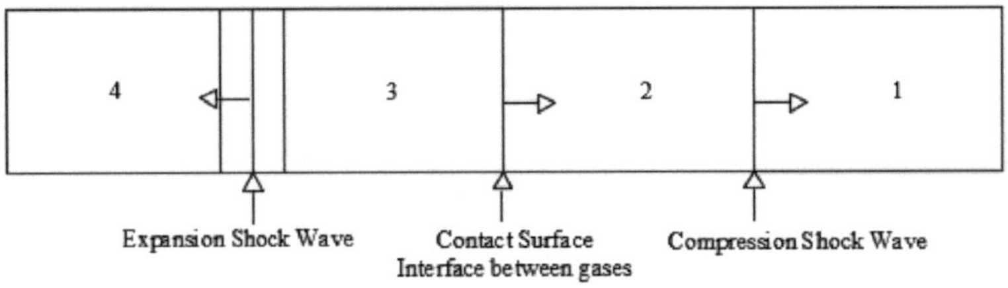

Figure 2.
Contact surfaces and four regions formed after the diaphragm rupture (adapted from [3]).

The incident shock wave is reflected and propagates towards the driver section when it reaches the closed end of the shock tube driven section. The reflected shock wave superimposes the motion of the incident shock wave, this superposition increases temperature and pressure in the driven section. The reflected shock wave is responsible for causing the dissociation and ionization of the gas inside the shock tube, [2, 4]. The **Figure 3** shows the propagation shock wave, reflected wave, expansion wave, reflected expansion wave and the contact surface after the diaphragm rupture shock tube.

The shock force is determined by the pressure ratio (P4/P1) and speed of sound propagation (a4/a1) between the driver and driven sections, [3]. The **Figure 4** shows the conditions of pressure and temperature in the driver and driven sections before the diagram rupture.

In the front of the shock and expansion waves the pressure, density and temperature do not varied in relation to the initial conditions in the driver and driven

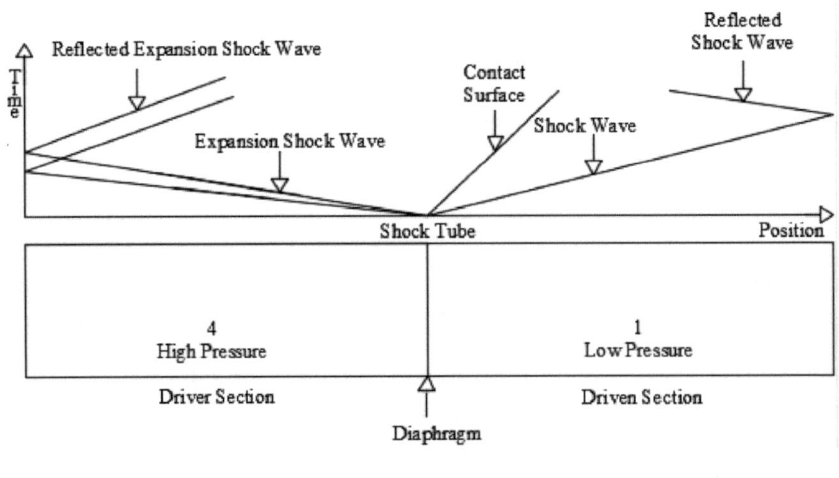

Figure 3.
Shock wave, reflected wave, expansion wave, reflected expansion wave and the contact surface after the diaphragm rupture shock tube.

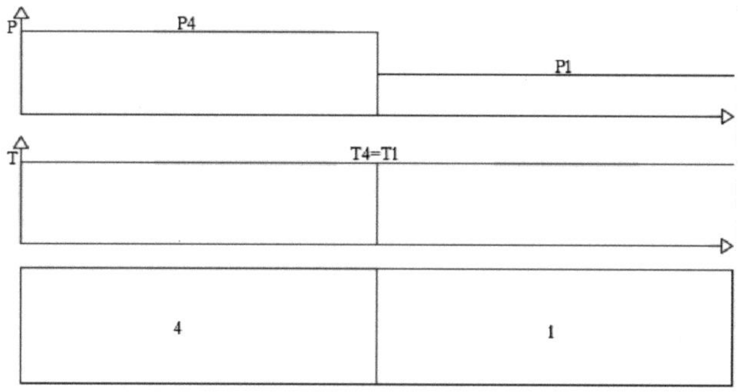

Figure 4.
Condition of pressure and temperature in the driver and driven sections before the diagram rupture (adapted from [3]).

sections, these sections are not affected by the passage of the shock and expansion waves. Behind the shock wave, pressure, density and temperature increase, while behind the expansion wave these variables decrease, [3]. The **Figure 5** shows the conditions of pressure and temperature in the driver and driven sections after the diagram rupture.

The **Figure 6** shows the conditions of shock tube after reflection shock wave and reflection expansion wave.

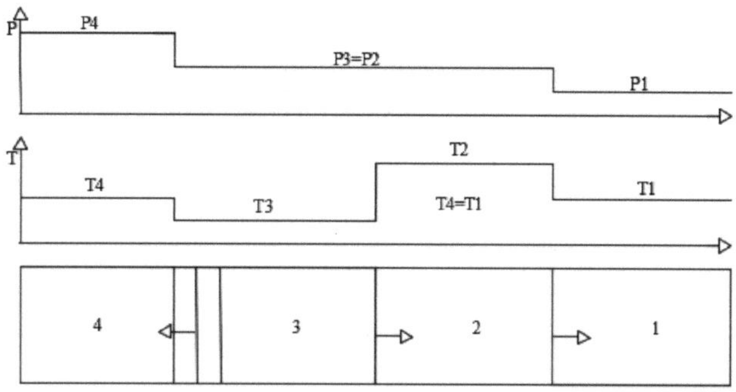

Figure 5.
Condition of pressure and temperature in the driver and driven sections after the diagram rupture (adapted from [3]).

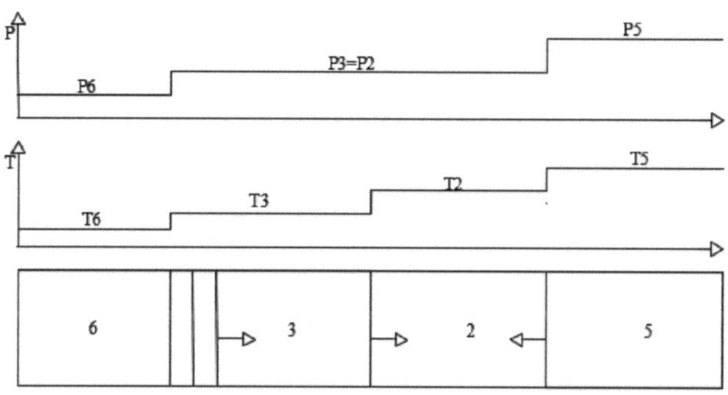

Figure 6.
Conditions of shock tube after reflection shock wave and reflection expansion wave (adapted from [3]).

2. Analytical solution shock tube for ideal gas

The speed of sound for each state gas must be calculated using the Eq. (1).

$$a = \sqrt{\gamma RT} \tag{1}$$

Where γ is the ratio of specific heats of the gas, *R* is the universal gas constant and *T* is the gas temperature in the respective regions of the shock tube. The Mach number can be determined using the Eq. (2), [3].

$$\frac{P_4}{P_1} = \frac{\gamma_{1-1}}{\gamma_{1+1}}\left[\frac{2\gamma_1}{\gamma_{1-1}}M_s^2 - 1\right]\left[1 - \frac{\frac{\gamma_{4-1}}{\gamma_{4+1}}\left(\frac{a_1}{a_4}\right)\left(M_s^2 - 1\right)}{M_s}\right]^{-\frac{2\gamma_4}{\gamma_{4-1}}} \tag{2}$$

Where *Ms* is the shock wave Mach number, the subscript 1 denotes the properties of the driven section and the subscript 4 denotes the properties of the driver section. The pressure ratio (P2/P1) on both sides of the shock wave can be calculated using the Eq. (3), [3].

$$\frac{P_2}{P_1} = 1 + \frac{2\gamma_1}{\gamma_1 + 1}\left(M_s^2 - 1\right) \tag{3}$$

The pressure ratio (P2/P1) can be used to determine the temperature ratio (T2/T1) on both sides of the shock wave using the Eq. (4), [3].

$$\frac{T_2}{T_1} = \frac{P_2}{P_1}\left(\frac{\frac{\gamma_1+1}{\gamma_1-1} + \frac{P_2}{P_1}}{1 + \frac{\gamma_1+1}{\gamma_1-1}\frac{P_2}{P_1}}\right) \tag{4}$$

The shock wave reflected Mach number M_R depend the velocity of the incident shock wave and can be calculated by the Eq. (5), [3].

$$\frac{M_R}{M_R^2 - 1} = \frac{M_s}{M_s^2 - 1}\sqrt{1 + \frac{2(\gamma_1 - 1)}{(\gamma_1 + 1)^2}\left(M_s^2 - 1\right)\left(\gamma_1 + \frac{1}{M_s^2}\right)} \tag{5}$$

The increased pressure of the shock wave reflected *P5* depend the speed of the incident shock wave, this ratio can be calculated by the Eq. (6), [3].

$$\frac{P_5}{P_2} = 1 + \frac{2\gamma_1}{\gamma_1 + 1}\left(M_R^2 - 1\right) \tag{6}$$

The Eq. (7) shows the relationship of the motion of the mass gas behind the shock wave and the reflected shock wave, [3].

$$\frac{2a_1}{\gamma_1 + 1}\left(M_s - \frac{1}{M_s}\right) = \frac{2a_2}{\gamma_1 + 1}\left(M_R - \frac{1}{M_R}\right) \tag{7}$$

The calculations involving speed of the gas molecules can be determined by the Eq. (8), [3].

$$Mach\ M = \frac{V}{a} \tag{8}$$

The calculation of the speed of the shock wave can be determined by the Eq. (9), [3].

$$V_R = M_R a_2 - V_2 \tag{9}$$

To determine the relationship between reflected shock wave pressure and incident shock wave pressure using the Eq. (10), [3].

$$\frac{P_5}{P_2} = \left(\frac{\frac{\gamma_1+1}{\gamma_1-1} + 2 - \frac{P_1}{P_2}}{1 + \frac{\gamma_1+1}{\gamma_1-1}\frac{P_1}{P_2}} \right) \tag{10}$$

With the ratio compression known the ratio reflected shock wave temperature and incident shock wave temperature can be determined by Eq. (11), [3].

$$\frac{T_5}{T_2} = \frac{P_5}{P_2} \left(\frac{\frac{\gamma_1+1}{\gamma_1-1} + \frac{P_5}{P_2}}{1 + \frac{\gamma_1+1}{\gamma_1-1}\frac{P_5}{P_2}} \right) \tag{11}$$

The temperature and pressure behind the reflected shock wave can be calculated knowing only the Mach number of the incident shock wave. This value can be determined from the velocity of the gas driven and wave velocity, [3].

3. Works carried out in shock tube

[5] used a shock tube to measure the ignition delay times of mixture with ethanol, n-heptane and iso-octane and mixture with ethanol, iso-octane, n-heptane and toluene. The tests were performed at temperatures ranging from 690 to 1200 K and pressures at 10, 30 and 50 bar. For testing mixture with ethanol, n-heptane and iso-octane were found delay times ranging from 120 to 6230 microseconds, for testing mixture with iso-octane, toluene, n-heptane and ethanol were found delay times ranging from 28 to 8731 microseconds and for testing mixture with iso-octane, toluene and n-heptane were found ignition delay times ranging from 180 to 1060 microseconds. [6] also used a shock tube to measure the ignition delay times of mixture with n-heptane and n butanol. The tests were performed at temperatures ranging from 1200 to 1500 K, pressures at 2 and 10 atm and equivalence ratios at 0.5 and 1. For testing with pure n-heptane were found delay times ranging from 90 to 1230 microseconds, for testing mixture with pure n-butanol were found delay times ranging from 120 to 950 microseconds and for testing mixture with n-heptane and n-butanol were found ignition delay times ranging from 30 to 1010 microseconds. [7] also conducted shock tube tests with n-heptane, iso-octane and ethanol. The tests were performed at temperatures ranging from 690 to 1200 K and pressures at 10, 30 and 50 bar. For testing at 10 bar and mixture with n-heptane, iso-octane and ethanol were found ignition delay times ranging from 181 to 2870 microseconds. For testing at 30 bar and mixture with n-heptane, iso-octane and ethanol were found ignition delay times ranging from 172 to 7800 microseconds. For testing at 50 bar and mixture with n-heptane, iso-octane and ethanol were found ignition delay times ranging from 115 to 7690 microseconds. For testing at 10 bar and mixture with n-heptane, iso-octane, toluene and di-isobutylene were found ignition delay times ranging from 245 to 4600 microseconds. For testing at 30 bar and mixture with n-heptane, iso-octane, toluene and di-isobutylene were found ignition delay times ranging from 191 to 8320 microseconds. For testing at 10 bar and mixture with n-heptane, iso-octane, toluene and di-isobutylene were

found ignition delay times ranging from 149 to 10100 microseconds. [8] also conducted shock tube tests with n-heptane and were found ignition delay times ranging from 1220 to 10600 microseconds. The tests were performed at temperatures ranging from 651 to 823 K, pressures at 6.1 and 7.4 atm and equivalence ratio of 0.75. [9] also conducted shock tube tests with propane and were found ignition delay times ranging from 100 to 11000 microseconds for testing at 6 atm. Were found ignition delay times ranging from 200 to 11000 microseconds for testing at 24 atm and were found ignition delay times ranging from 300 to 600 microseconds for testing at 60 atm. The tests were performed at temperatures ranging from 980 to 1400 K and equivalence ratio 0.5. [10] also conducted shock tube tests with methyl butanoate and were found ignition delay times ranging from 19630 to 24180 microseconds for testing at 10.2 atm. The tests were performed at temperatures from 985 K and equivalence ratio 0.3. [11] also conducted shock tube tests with methyl octanoate, n-nonane and methylcyclohexane. The tests were performed at temperatures ranging from 1263 to 1672 K, pressures at 1.5 and 10 atm and equivalence ratio 0.5, 1 and 2. For tests with equivalence ratio 0.5 at 1.5 atm were found for methyl octanoate ignition delay times ranging from 40 to 1000 microseconds, for n-nonane were found delay times ranging from 100 to 1100 microseconds and for methylcyclohexane were found delay times ranging from 100 to 1200 microseconds. For tests with equivalence ratio 0.5 at 10 atm were found for methyl octanoate ignition delay times ranging from 110 to 800 microseconds, for n-nonane were found delay times ranging from 90 to 900 microseconds and for methylcyclohexane were found delay times ranging from 90 to 1050 microseconds. For tests with equivalence ratio 1 at 1.5 atm were found for methyl octanoate ignition delay times ranging from 120 to 1000 microseconds, for n-nonane were found delay times ranging from 90 to 1100 microseconds and for methylcyclohexane were found delay times ranging from 120 to 1100 microseconds. For tests with equivalence ratio 1 at 10 atm were found for methyl octanoate ignition delay times ranging from 80 to 1000 microseconds, for n-nonane were found delay times ranging from 80 to 1100 microseconds and for methylcyclohexane were found delay times ranging from 150 to 1100 microseconds. For tests with equivalence ratio 2 at 1.5 atm were found for methyl octanoate ignition delay times ranging from 100 to 900 microseconds, for n-nonane and for methylcyclohexane were found delay times ranging from 100 to 1100 microseconds. For tests with equivalence ratio 2 at 10 atm were found for methyl octanoate and methylcyclohexane ignition delay times ranging from 90 to 1000 microseconds, for n-nonane were found delay times ranging from 100 to 1100 microseconds. [12] also conducted shock tube tests with methyl stearate, methyl oleate, methyl linoleate, methyl linolenate, and methyl palmitate and were found ignition delay times for all fuel testing ranging from 200 to 90000 microseconds for testing at 13.5 bar. The tests were performed at temperatures from 700 to 1100 K and equivalence ratio 1. [13] also conducted shock tube tests with jets fuels, rocket propellants, diesel fuel and gasoline fuel and were found ignition delay times for all fuel testing ranging from 100 to 1900 microseconds for testing at pressure from 6 to 60 atm. The tests were performed at temperatures from 1000 to 1400 K and equivalence ratio 0.85 and 1.15. [14] also conducted shock tube tests with mixture of biodiesel with diesel fuel and were found ignition delay times for all fuel testing ranging from 60 to 2600 microseconds for testing at pressure at 0.12 Mpa. The tests were performed at temperatures from 1174 to 1685 K and equivalence ratio 0.5, 1 and 1.5. [15] also conducted shock tube tests with diesel fuel and alternative hydro processed jet fuels. The tests were performed at temperatures from 650 to 1300 K, pressures from 0.8 to 80 atm and equivalence ratio 0.25 to 1.5. For testing with jet fuels were found ignition delay times ranging from 60 to 8000 microseconds and for testing with diesel fuels were found ignition delay times ranging from 90 to

4000 microseconds. [16] also conducted shock tube tests with conventional and alternative jet fuels, alcohol to jet, direct sugar to hydrocarbon, biodiesel-like fuel, n-heptane, n-dodecane, m-xylene and iso-dodecane. Were found ignition delay times for all fuel testing ranging from 20 to 3200 microseconds. The tests were performed at temperatures from 980 to 1800 K, pressures at 16 atm and equivalence ratio at 0.5.

4. Experimental measuring the ignition delay times of the convectional diesel and biodiesel from soybean oil using a shock tube

The experiments were conducted in the heated shock tube facility of the Mobility Technology Center (CTM) of Federal University of Minas Gerais (UFMG). The shock tube has a 3 m long driver section and a 3 m long driven section with an internal diameter of 97.20 mm. Aluminum diaphragm of 0.4 mm thickness divided the driver and driven sections before each experiment. The experiments were carried out with convectional Diesel and pure biodiesel from soy oil. The convectional Diesel used in this study is normally fuel found at gas stations and it has a cetane number of 43. The pure biodiesel used in this study was derived from a process of refining oil from soy oil and it has a cetane number of 38. The instrumentation within the shock tube used for the experiment included three pressure sensors (P1, P2 and P3), two temperature sensors (T3 and T1), a luminosity detection sensor (L1) and a fuel injector (FI). In the present work a mixture of the Nitrogen (N_2) and Argon (Ar) gases was used as the driver gas to obtain a longer test time. The **Figure 7** shows the position of the sensors, fuel injector, aluminum diaphragm location and mixture of the Nitrogen and Argon inlet in the shock tube.

The pressure sensor P3 (located at 1700 mm before the aluminum diaphragm) was used to monitor the pressure in the driver section and the diaphragm rupture pressure. The pressure sensor P2 (located at 700 mm after the aluminum diaphragm) was used to indicates the moment of passage of the shock wave after diaphragm rupture. This information is used to control and define the fuel injection timing. The pressure sensor P1 (located at 2700 mm after the diaphragm) was used to indicates the moment of passage of shock in region 1 where combustion occurs. For monitor the temperature in the driver section was used an analog temperature sensor T3. For monitor and control the temperature in the driven section was used a temperature sensor T1 with the same characteristics that the temperature sensor T3 used in driver section. The luminosity detection sensor L1 was used to indicates the

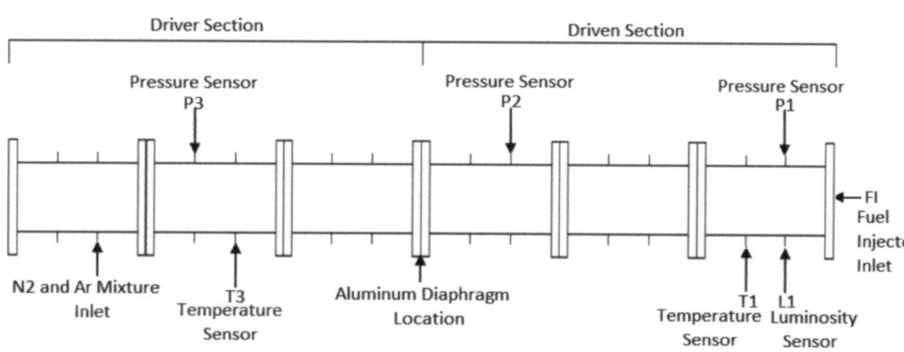

Figure 7.
Position of the sensors, fuel injector, aluminum diaphragm location and mixture of the nitrogen and argon inlet in the shock tube (adapted from [17]).

moment of combustion. At the moment of ignition, the voltage of this sensor decreases in function of flame in shock tube. This information together with the pressure signal of P1 sensor was used to calculated the ignition delay time. The fuel injector FI injects fuel into the shock tube when the sensor pressure P2 detects the

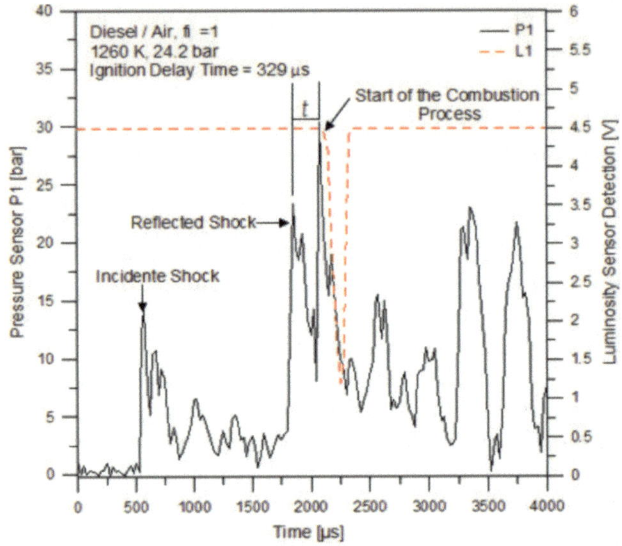

Figure 8.
Ignition delay time of diesel in air at shock reflected pressure of 24.2 bar, equivalence ratio of 1 and temperature of 1260 K (adapted from [17]).

Incident shock		Reflected shock		Ignition delay time
P2 (bar)	**T2 (K)**	**P5 (bar)**	**T5 (K)**	**τ (μs)**
14.2	932	24.6	1150	362
14.2	945	24.3	1162	342
14	962	24.2	1260	329
13.7	874	23.6	940	603
14.2	912	23.8	1065	316
14.1	918	24.6	1082	418
13.9	915	23.1	980	443
13.4	902	24.2	1008	439
13.2	874	24.7	972	518
13.9	862	24.3	965	780
13.6	854	24.7	903	856
14	862	24.1	920	790
14.2	840	24.2	972	680
13.9	798	24.8	995	648
13.8	823	24.5	1040	490
13.6	890	23.8	1120	412
14	944	23.9	1243	325

Table 1.
Measured ignition delay times for convectional diesel in air (adapted from [17]).

passage of shock wave. The ignition delay time was calculated by the time difference between the passage of the shock wave by the P1 sensor and the start of the ignition detected by the luminosity detection sensor L1. The **Figure 8** shows a result of calculation of the ignition delay time of diesel measured in shock tube in air at shock reflected pressure of 24.2 bar, equivalence ratio of 1 and temperature of 1260 K.

The **Table 1** listed the measured ignition delay time τ for convectional diesel. All measurements were carried out equivalence ratios 1. The experiments were performed in the temperature range of 903 to 1260 K and target pressures were approximately 24 bar. Were found ignition delay times ranging from 316 to 856 microseconds.

Incident shock		Reflected shock		Ignition delay time
P2 (bar)	T2 (K)	P5 (bar)	T5 (K)	τ (μs)
14.6	942	24.7	1125	1356
14	918	24.4	960	1636
14.1	965	24.2	1150	945
13.2	932	23.6	995	1635
14.7	885	24.1	940	1525
14.3	948	24.5	1060	1567
13.9	935	23.2	1095	1470
13.4	842	24.7	916	1782
13.7	880	24.5	926	1678
14.7	982	24.8	1210	640
14.8	947	24.7	1180	982

Table 2.
Measured ignition delay times for pure biodiesel from soybean oil in air (adapted from [17]).

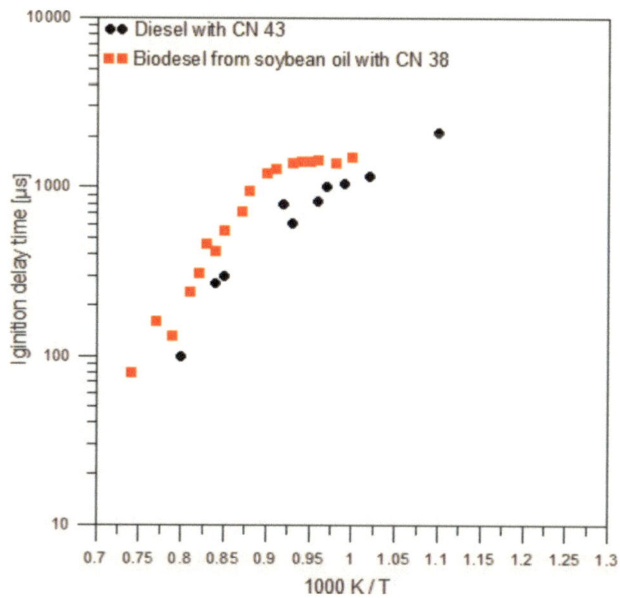

Figure 9.
Ignition delay times for diesel and biodiesel from soybean oil in present study (adapted from [17]).

The **Table 2** listed the measured ignition delay time τ for pure biodiesel from soybean. All measurements were carried out equivalence ratios 1. The experiments were performed in the temperature range of 916 to 1210 K and target pressures were approximately 24 bar. Were found ignition delay times ranging from 640 to 1782 microseconds.

The **Figure 9** compare the ignition delay times for diesel and biodiesel from soybean oil measured in present study.

The ignition delay times measured in this study are consistent with those found in the literature. For convectional Diesel fuel were found times ranging from 316 to 856 microseconds and for biodiesel from soybean oil fuel were found times ranging from 640 to 1782 microseconds. These measured values are consistent with expectations since convectional Diesel has a higher cetane number than biodiesel from soybean oil. Convectional Diesel have a cetane number 43 while biodiesel from soybean oil have cetane number 38. This study confirm that the ignition delay time decreases with increasing cetane number. Fuel with a high cetane number has a short ignition delay time and the best quality in the combustion process.

Author details

Claudio Marcio Santana[1] and Jose Eduardo Mautone Barros[2*]

1 Federal University of Ouro Preto, Ouro Preto, Brazil

2 University Federal of Minas Gerais, Belo Horizonte, Brazil

*Address all correspondence to: mautone@demec.ufmg.br

References

[1] Sarathy, S. M., Farooq, A. and Kalghatgib, G. T., **Recent progress in gasoline surrogate fuels,** Progress in Energy and Combustion Science 65 (2018) 67–108, doi.org/10.1016/j.pecs.2017.09.004.

[2] Glass, I. I., **Shock Tubes**: Part I. Toronto: University of Toronto, 1958.

[3] Mcmillan, R.J. **Shock tube investigation of pressure and ion sensors used in pulse detonation engine research,** (2012), Corpus ID: 107883156.

[4] H. J. Gordon, **Shock Tubes**: Part II. Toronto: University of Toronto, 1958.

[5] L. R. Cancino, M. Fikri, A. A. M. Oliveira, C. Shulz, **Autoignition of gasoline surrogate mixtures at intermediate temperatures and high pressures: Experimental and numerical approaches,** Proceeding of the Combustion institute, 32 (2009) 501–508, doi:10.1016/j.proci.2008.06.180.

[6] S. Niu, J. Zhang, Y. Zhang, C. Tang, X. Jiang, E. Hu, Z. Huang, **Experimental and modeling study of thc auto-ignition of n-heptane and n-butanol mixtures,** Combustion and Flame 160 (2013) 31–39, doi:10.1016/j.combustflame.2012.09.006.

[7] M. Fikri, J. Herzler, R. Starke, C. Schulz, P. Roth, G. T. Kalghatgi, **Autoignition of gasoline surrogates' mixtures at intermediate temperatures and high pressures,** Combustion and Flame 152 (2008) 276–281. doi:10.1016/j.combustflame.2007.07.010.

[8] M. F. Campbell, S. Wang, C. S. Goldenstein, R. M. Spearrin, A. M. Tulgestke, L. T. Zaczek, D. F. Davidson, R. K. Hanson, **Constrained reaction volume shock tube study of *n*-heptane oxidation: Ignition delay times and time-histories of multiple species and temperature,** Proceedings of the Comb ustion Institute 35 (2015) 231–239. doi: 10.1016/j.proci.2014.05.001.

[9] K. Y. Lam, Z. Hong, D. F. Davison, R. K. Hanson, **Shock tube ignition delay time measurements in propane / O_2 / argon mixtures at near-constant-volume conditions,** Proceeding of the Combustion Institute 33 (2011) 251–258.

[10] S. M. Walton, D. M. Karwat, P. D. Teini, A. M. Gorny, M. S. Wooldridge, **Speciation studies of methyl butanoate ignition,** Fuel 90 (2011) 1796–1804.

[11] B. Rotavera, E.L. Petersen, **Ignition behavior of pure and blended methyl octanoate, n-nonane and methylcyclohexane,** Proceeding of the Combustion Institute 34 (2013) 435–442.

[12] C. K. Westbrook, C. V. Naikb, O. Herbinetc, W. J. Pitza, M. Mehla, S. M. Sarathya, H. J. Currand, **Detailed chemical kinetic reaction mechanisms for soy and rapeseed biodiesel fuels,** Combustion and Flame 158 (2011) 742–755, doi: 10.1016/j.combustflame.2010.10.020.

[13] D.F. Davidson, Y. Zhu, J. Shao, R.K. Hanson, **Ignition delay time correlations for distillate fuels,** Fuel 187 (2017) 26–32. doi.org/10.1016/j.fuel.2016.09.047.

[14] V. N. Hoang, L. D. Thi, **Experimental study of the ignition delay of diesel and biodiesel blends using a shock tube,** Biosystems engineering 134 (2015) 1–7. doi.org/10.1016/j.biosystemseng.2015.03.009.

[15] S. Gowdagiri, M. A. Oehlschlaeger, **Global Reduced Model for Conventional and Alternative Jet and**

Diesel Fuel Autoignition, Energy Fuels 28 (2014) 2795–2801. doi.org/10.1021/ef500346m.

[16] G. Flora, J. Balagurunathan, S. Saxena, J. P. Cain, M. S. P. Kahandawala, M. J. DeWitt, S. S. Sidhua, E. Corporan, **Chemical ignition delay of candidate drop-in replacement jet fuels under fuel-lean conditions: A shock tube study,** Fuel 209 (2017) 457–472. doi.org/10.1016/j.fuel.2017.07.082.

[17] Santana, C.S., Barros, J. E. M., Junior, H. A. A, Braga, J. O. and Neto, J. C. B, **Measuring and comparing the ignition delay time of the reference diesel, convectional diesel, additive ethanol and biodiesel from soybean oil using a shock tube,** Journal of the Brazilian Society of Mechanical Sciences and Engineering (2020) 42:102, doi: 10.1007/s40430-020-2183-z.

Chapter 5

Assessing the Effects of Engine Load on Compression Ignition Engines Using Biodiesel Blends

Semakula Maroa and Freddie Inambao

Abstract

This study evaluated the performance of a diesel engine operated with waste plastic biodiesel fuel (WPPO) blends. Findings were that at all engine loads (from idling to full load) the emissions of carbon monoxide (CO), unburnt hydrocarbon (UHC) and carbon dioxide (CO_2) were low compared to conventional diesel (PD), although the emissions of NO_X were higher. The brake specific fuel consumption (BSFC) for the blends dropped while the brake thermal efficiency (BTE) increased with load for all blends until intermediate load when it decreased. WPPO blends had a higher viscosity compared to PD. CO emissions for blend 95/WPPO5 at all engine speed idling modes were 285 ppm, 298 ppm, 320ppm, and 388 ppm while PD emissions were 270 ppm, 295 ppm, 315 ppm and 365 ppm respectively. The values for UHC for blend 95/WPPO5 at all modes were 35 ppm, 28 ppm, 22 ppm, and 18 ppm compared to PD fuel with 20ppm, 25 ppm, 30 ppm, and 40ppm respectively. The NO_X emissions for PD fuel at all modes were 175 ppm, 225 ppm, 300 ppm and 375 ppm compared to blend 95/WPPO5 at 195 ppm, 245 ppm, 335 ppm, and 397 ppm. The BSFC values for blend 95/WPPO5 at all modes were 0.48 kg/kW.h, 0.41 kg/kW.h, 0.35 kg/kW.h and 0.4 kg/kW.h compared to PD at 0.45 kg/kW.h, 0.39 kg/kW.h, 0.33 kg/kW.h and 035 kg/kW.h respectively.

Keywords: engine loads, emissions, higher viscosity, spray characteristics

1. Introduction

Development of alternative fuel energy began in the 1900s when German engineer Rudolf Diesel invented the diesel engine using vegetable oil [1]. However, due to availability of petroleum at the time the focus moved into fossil fuel to the disadvantage of bio-oil. Currently many researchers such as [2–7] have focused on development of alternative fuel to petro-diesel (PD). Most of this research is heavily biodiesel based as this is one of the solutions to replace fossil fuels while creating renewable and green fuels. Fossil fuels are non-renewable and are depleting rapidly, hence the need for large-scale research to find alternative and renewable fuels. Alternative fuels must prove to be feasible, environmentally friendly and sustainable while meeting a large energy demand [8].

Fossil fuels have a detrimental environmental impact [9] when released to the atmosphere due to the combustion activities of fossil fuels. It is being projected that if no measures are put in place by 2030 the use of fossil fuel will raise emission levels by 39% [10]. Besides environmental concerns, fossil fuels have erratic demand and supply which increases international market prices and other commodities hence promoting inflation [11]. **Figure 1** shows measures taken as a way of combating environmental pollution from the transportation industry in the European Union by means of taxes and social contribution as a function of gross domestic product.

In order to determine the efficacy of biodiesel, mainstream researchers in biodiesel fuels have evaluated engine performance using different feedstocks and different biodiesel blends [13–16]. However, few have been able to investigate the influence of load using plastic waste oil blends of biodiesel [17–19]. All these researchers have concentrated on performance and emission characteristics with little attention to low load and intermediate load compared to engine full load [8, 20, 21]. For example, all low load and intermediate engine idling speeds are considered as high idling, and mostly increase emissions from trucks and vehicles in the transport industry.

High idling or low engine loads have been shown to increase NO_X emissions on roads compared to high speed road driving by a factor of 1.5 [22, 23]. In other words, increasing low load increases NO_X emissions [24–26]. During idling, which is low load, the fuel consumption as well as engine wear and maintenance increase. The average fuel consumption for example in trucks at idle is 0.8 g/hr. to 1.5 g/hr. based on the size of the engine, ambient temperature and the load of other systems such as HVAC and vehicle electrical loads [27]. This is when compared to driving cycle emissions of UHC that are 1–5 times more.

On the other hand, during low load other emissions such as CO rise to 295 g/hr. [28–30]. The carbon emissions during the driving cycles are estimated at 45–75%, while UHC emissions during idling and low load can reach 86.4 g/hr. [27, 31]. Most diesel engines typically spend a substantial amount of time in idling mode, either at traffic stops, checkpoints or in exchange periods in fuel stations. The idle time spent varies considerably with the many varied reasons for maintaining engines at idle. For long haulage trucks, for example, the most common reason is climate control, loading and offloading transport cargo or service and maintenance [32, 33]. The other reason why trucks idle for a long time is use of the engines to heat and air-condition cabs and to power amenities in the cab while on the road [34, 35].

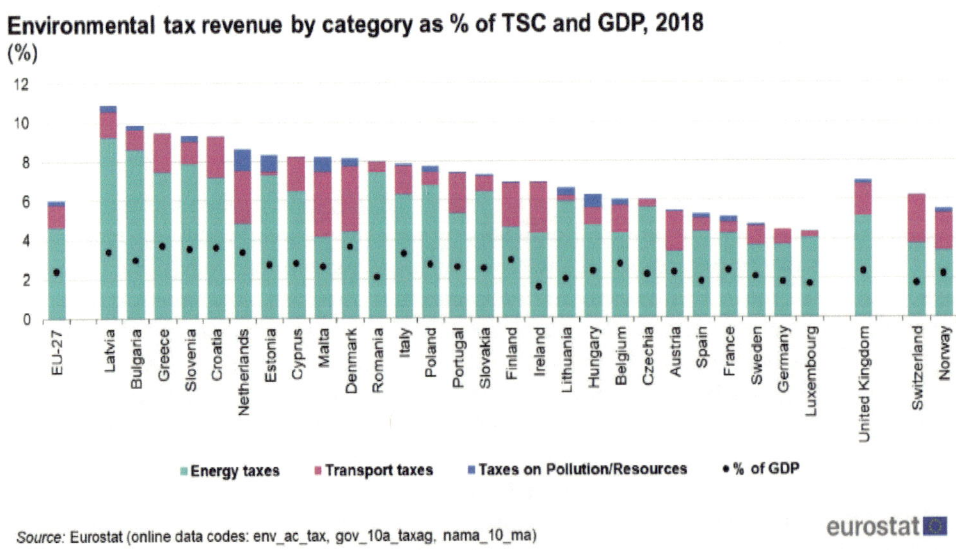

Figure 1.
Environmental taxes as % of GDP and as % of total taxes and social contributions [12].

Biodiesel oil is known to contain physicochemical characteristics of functional PD properties [36, 37]. Research has shown biodiesel fuels have many advantages over PD. For example, biodiesels are biodegradable, non-flammable, renewable, non-explosive, non-toxic and environmentally friendly [38, 39]. These qualities show biodiesel fuels to be the best options to substitute for fossil fuels. Biodiesel fuels have a variety of feedstocks such as used vegetable oil, waste plastics, waste biomass, animal fats (tallow) and recently microalgae, all which can be processed into biodiesel [40]. Biodiesel has the ability to be utilized as a fuel with or without engine modification which gives it high technical advantage [41, 42].

The use of biodiesel and biodiesel blends affect diesel engine performance characteristics. Poor quality biodiesel fuel results in deposits and clogging [43, 44]. Besides these problems, use of biodiesel can result in corrosion, excessive engine wear and premature engine failure [45]. Biodiesel also causes deposits in the injector pump, which interferes with the spray pattern, an essential factor in mixing fuel during the combustion process, hence poor engine performance [46]. Other demerits, which are associated with biodiesel fuel use, include dilution of lubrication oil leading to high crank-case oil levels followed by loss of engine oil pressure and increased engine bearing wear. Thus, it is clear that the quality and testing of biodiesel is an important factor in ensuring proper rating, acceptance and durability of diesel engines.

The objective of this work was to use waste plastic pyrolysis oil (WPPO) and determine the effects of idling speed load-using blends of WPPO on a diesel engine. The second objective was to study the effect of brake specific fuel consumption (BSFC) of WPPO at low and intermediate engine conditions, also known as high idling condition. The third objective was to find the effect of engine load at high idling on engine performance and emission characteristics using WPPO as an alternative fuel.

2. Methodology and materials

2.1 Crude WPPO oil properties

WPPO was selected for this study because of the advantage of turning waste into energy to reduce the environmental impact of waste plastic. The second factor that informed the use of waste plastic is sustainability as waste plastic is readily available in municipal solid waste management sites. The plastics were collected from various holding facilities within the Durban metropolitan centres comprising a variety of plastics.

The pyrolysis oil was obtained from the pyrolysis unit in the Green Energy Group laboratory in the Department of Mechanical Engineering, University of KwaZulu-Natal. The author in his previous work covered the design of the unit and its performance analysis was published in the proceedings of the DUE 2019 conference in Cape Town [47]. The WPPO testing and measurements were conducted at InterTek, a private laboratory in Durban and the results are shown in **Table 1**.

2.2 WPPO biodiesel processing

A two-step process was used to process the WPPO as its acid value is higher compared to petroleum diesel. Therefore, an acid catalyzed process was used with the molar ratio maintained at 12: (50% v. v), 1% of H_2SO_4 was added to the preheated oil at 70°C for 3.5 hrs with a stirring speed of 400 r/min in a reactor of 5 liters.

	Unit	PD	WPPO
Density @ 20 °C	Kg/M3	845	825
K. Viscosity @ 40 °C	mm^2/s	3.04	2.538
Cetane number	—	55	—
Flash point	°C	50	43
Fire point	°C	56	45
Carbon residue	%	22	0.015
Sulfur	%	<0.028	0.248
Gross calories	MJ/kg	46.50	43.32

Table 1.
Properties of diesel and WPPO before processing into biodiesel properties.

Thereafter the products were put into a separating funnel and the excess alcohol, sulfuric acid and other impurities in the upper layer were drained.

To remove methanol and water from the esterified oil a rotary evaporator was employed at 100°C under vacuum for 1 hour and 20minutes.

To complete the process reaction an alkaline catalyzed process was employed by reacting the esterified oil with methanol at 6:1 molar ratio and 1% potassium hydroxide (KOH) at 80°C for 2 hours and a stirring speed of 400 r/min.

The final step to obtain a refined biodiesel oil was to leave the produced biodiesel in a separation funnel overnight, for the reaction to end. This process required 12 hours to finish reacting before the lower layer of impurities can be discarded.

2.3 WPPO fatty acid composition

The fatty acid for a double bond is unsaturated, so a single bond fatty acid, which is saturated, was tested using the FT-IR and confirmed by the GC–MS method. **Table 2** shows the GC–MS operating conditions while **Table 3** shows the FT-IR indicated compounds of pyrolysis biodiesel oil and their class of compound.

The biodiesel obtained was composed of more than 20 compounds of mixed proportion whose composition and GC–MS percentage areas spectrum are presented in **Table 4**. **Table 5** has a list of test equipment utilized in the experiment.

Considering percentage areas of the spectrum, the highest pick areas of the total chromatography were the following: heptadecane, n-octadecane, n-hexadecane, nonadecane, pentadecane, eicosane, tetradecane and tridecane. Eq. 1 shows the effect of linear velocity of the carrier gas in retention time which was used to determine the carrier gas linear velocity.

$$t_r = L\frac{(K+1)}{\mu}\# \quad (1)$$

Where.

t_r is the retention time.

L is the column height.

K is the retention factor (constant).

μ is the carrier gas linear velocity.

The components present in mixed WPPO range from carbon number C_{10} to C_{40}. A large percentage of these components are made of aliphatic compounds as shown by the result the GC–MS spectrum result in **Table 4**.

Property	Specification
Carrier gas	Helium @ 23.8 psi
Linear velocity	44 cm/s@100°C
Flow rate	Air = 450 ml/min
	H2 = 40 ml/min
	He = 20 ml/min
Injector	Split injector, 50:1ratio, 0.3 μL injection volume
Temperature ramp 1	100°C zero minutes hold
Temperature ramp 2	10 °C/min to 250°C 5 minutes hold
Detector temperature	250°C
Column head pressure	23.8

Table 2.
Showing GC–MS operating conditions during the experiment.

Frequency range (cm^{-1})	Group	Class compound
3750–3250	O-H stretching	Polymeric O-H, HO_2 impurities
3150–2950	C-H stretching	Alkanes
1950–1830	C=O stretching	Ketones, aldehydes, carboxylic acid
1830–1725	C ≡ C stretching	alkenes
1725–1575	$-NO_2$ stretching	Nitrogenous compounds
1575–1475	C-H bending	alkanes
1475–1375	C-O stretching	Primary/secondary alcohols
1325–1200	O-H bending	Esters, ethers, phenols
1175–1150	C-H bending	alkanes
1000–950	C ≡ C stretching	alkynes
900–875	—	Aromatic compounds

Table 3.
FT-IR WPPO indicated compounds of pyrolysis biodiesel oil.

Composition	Chemical name	Percentage
C10	Aliphatic compounds	65
C10-C13	Doxosane	2.4
C13-C16	Isoparaffin	7.5
C16 - C20	1-hexadecene	3.1
C20 – C23	Eicosane	7.6
C23-C30	Docosane	15.4
C		81.5
H		11.3
O		7.2

Table 4.
Elemental fatty acid composition of WPPO.

Property	Equipment	Standard
Kinematic viscosity	SVM 4000 (Anton Paar, UK)	ASTM D445
Flash point	NPM 550 (Norma lab, France)	ASTM D93
Oxidation stability	900 Rancimat (Metrohm, Switzerland)	ASTM D14112
CP/PP	NTE 500 (Norma lab, France)	ASTM D2500
Carbon residue	NMC 440 (Norma lab, France)	ASTM D4530
Total sulfur	5000 MULTI-EA (AJ Germany)	ASTM D5433
Calorific value	C 2500 basic calorimeter (IKA, UK)	ASTM D240
Density	SVM 3500 (Anton Paar, UK)	ASTM D1298

Table 5.
List of equipment used in the experiment.

2.4 WPPO properties analysis

In order to determine the physicochemical properties of the WPPO biodiesel Characterization tests were conducted based on the requirements and standards of ASTM D6751. Under this section, the following numbers were calculated using the fatty acid composition and empirical Equations [48, 49]. This included the saponification number, the cetane number and the iodine number. The saponification value is according to Eq. (2):

$$SN = \sum \frac{560 \times A_i}{MW_i} \# \quad (2)$$

The iodine value is according to Eq. (3):

$$IV = \sum \frac{254 \times D \times A_i}{MW_i} \# \quad (3)$$

The cetane index number is according to Eq. (4):

$$CN = 46.3 + \frac{5458}{SN} - (0.22 \times IV) \# \quad (4)$$

Where:
A_i is the weight percentage of each fatty acid component.
D is the number of double bonds in each fatty acid.
MW_i is the molecular weight.

To ensure proper mixing and blending of the various ratios during the experiment homogenous mixing equipment was used at speeds of 1800 r/min to 2000 r/min.

2.5 Engine testing and performance analysis

The engine test was conducted on a four-cylinder Iveco diesel dual fuel engine. To help in the analysis of the engine pressure, sensors and crankshaft position sensor and encoder were used. The aim of these two sensors was to provide the in-cylinder pressure in relation to the crankshaft position variation, using LabVIEW software. Combustion data was obtained, and graphs sketched.

The engine was coupled to a mechanical dynamometer with idling positions set at 500 r/min considered to be equivalent to 25% load, 1000 r/min for Mode 1

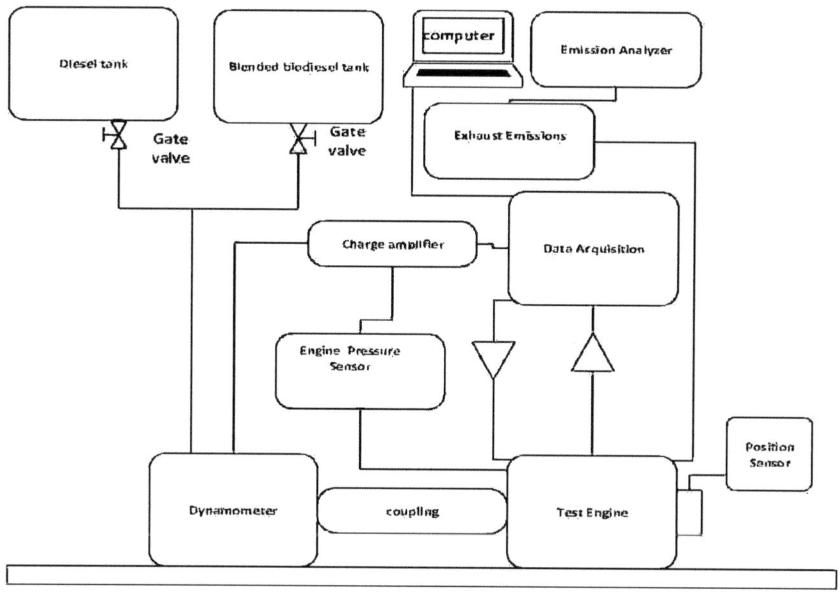

Figure 2.
Schematic diagram of the engine testing and equipment.

Parameters	Position value
Ignition type	4 (Stroke)DICI
Number of cylinders	4 in-line
Cooling medium	Water
Manufacturer	Iveco
Revolutions per minute	2000
Brake power	43.40 kW@2000
Cylinder bore	104 mm
Piston stroke	115 mm
Compression ratio	17:1
Connecting-rod length	234
Engine capacity	2500 cc
Dynamometer make	234
Injection timing	12°bTDC
Maximum torque	206.9 Nm @1500
Injection pressure	250–272 Bar

Table 6.
Experimental engine specifications.

equivalent to 50% engine load. For intermediate speeds two speeds are chosen as 1500 r/min and full load at 2000 r/min for Mode 2 as 75% and 100% engine load equivalents respectively. For engine load the dynamometer is fitted with a screw type loading device enabling each load to be synchronized with the intended engine speed targeted. **Figure 2** shows the schematic of the test engine and **Table 6** shows the engine specification.

3. Results and discussion

3.1 Brake specific fuel consumption (BSFC)

Fuel consumption as BSFC is a measure of fuel flow per unit time measured as a flow rate. The fuel thus measures how an engine utilizes supplied fuel to produce the intended work. While measuring BSFC lower values are preferred compared to higher values. The brake specific fuel consumption measures the efficiency of fuel by combustion of the fuel and air mixture, which does the actual work of crankshaft rotation. In other word, the BSFC is a ratio of the rate of fuel consumption in relation to the effective power produced by the engine. This means for every cycle of operation the BSFC tries to get an equal output with the corresponding increase in fuel supply to the engine (the engine is supplied volumetrically).

Figure 3 is a variation of BSFC with engine speed, and shows that as the speed increases, there is an equal increase in the fuel consumed by the test engine. The values obtained at full engine load for the blends of 95/WPPO5, 90/WPPO10, and 80/WPPO20, 70/WPPO30, 60/WPPO40 and PD were 0.4 kg/kW.h, 0.41 kg/kW.h, 0.42 kg/kW.h, 0.43 kg/kW.h and 0.35 kg/kW.h respectively. At high engine loads the conversion of heat energy to mechanical energy increased with increase in combustion temperature, leading to increased BSFC for the biodiesel. This increase was proportional to the difference in their heating values which is identical to the findings of [50]. Furthermore, the WPPO blends had high densities, therefore suffered high mass injection pressure, hence an increase in BSFC which is identical to studies by [51, 52]. WPPO blends compared well to conventional diesel fuel and sometimes other biodiesel blends with comparative differences in the heating values.

As the blend ratio increased there was a decrease in the BSFC across all the test fuels. However, the values for all WPPO blends increased compared to PD test fuel. This is due to the lower calorific values of the blends as the percentage of the blend ratio increased. In other words, by increasing the ratio of WPPO in the diesel test fuel the engine fuel consumption increased. This is identical to the studies of [53–55]. The closeness of the values and the packed graph reveal a close resemblance and identical BSFC characteristics of WPPO to PD properties. For example, **Figure 3** at Mode 1 (500 r/min (25% load) to 1000 r/min 50% engine load) blend 90/WPPO10 had a value of 0.48 kg/kW.h and 0.43 kg/kW.h compared to full engine speed Mode 2 2000 r/min

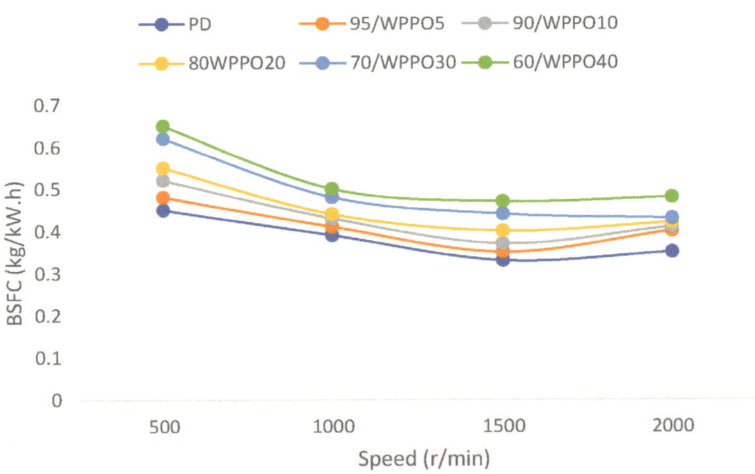

Figure 3.
Brake specific fuel consumption versus speed.

100% load) with 0.37 kg/kW.h and 0.41 kg/kW.h. This value is higher than PD with 0.4 kg/kW.h at 1000 r/min 50% engine load and 0.35 kg/kW.h at full engine load.

3.2 Brake thermal efficiency (BTE)

The BTE measures the ratio of the engine brake power to the heat of combustion supplied by the fuel consumed by the engine. The brake thermal efficiency also determines how well the engine converts the heat energy into actual mechanical energy. The BTE is influenced by engine design, type of fuel used and the engine application [56]. High engine load seems to increase BTE as can be seen in intermediate loads of 1000 r/min 50% engine load to 1500 r/min 75% engine load. When operating at part load the gross thermal efficiency of any engine falls to 28%, which translates to a rating of 22% down, from a full load thermal efficiency at 39.1%. Modern on-road diesel engines provide a 42% BTE at full load but waste almost 28% of all fuel used through exhaust gases.

The BTE variations with engine load is shown in **Figure 4**. The graphs show that as the load increased there was an increase in the BTE across all the test fuel blends of WPPO and PD. The result of this experiment shows that the BTE increased as the load increased, explained by the reduction in the heat loss as the engine power (more fuel) increased with load.

At Mode 1 (1000 r/min, 50% engine load) the values for blends 95/WPPO5, 90/WPPO10, 80/WPPO20, 70/WPPO30, 60/WPPO40 and PD were 22%, 21%, 20%, 18%, 16.5% and 22.5% respectively. As the blend ratio and engine idling load increased there was an increase in BTE across the blends of WPPO, but with a decrease in the BTE within the blends. For example, at Mode 1 (500 r/min, 25% engine load), 95/WPPO5 had values of 14%, 22%, 26.5% and 25.7% compared to 80/WPPO20 with 12.5%, 20%, 22.5% and 23% respectively.

This decrease in BTE within the blends is due to the presence of aromatic compounds in waste pyrolysis plastic oil, which require a lot of energy to break [18]. Another critical factor that could be contributing to lower BTE among blends of WPPO compared to PD fuel is the higher combustion temperature characteristics observed in WPPO fuel blends leading to high heat transfer losses [57]. The main factors causing reduction in the BTE with use of blends is their lower heating values, low air to fuel mixing (poor atomization of blends during injection), high viscosity, high biodiesel density, or higher BSFC [58].

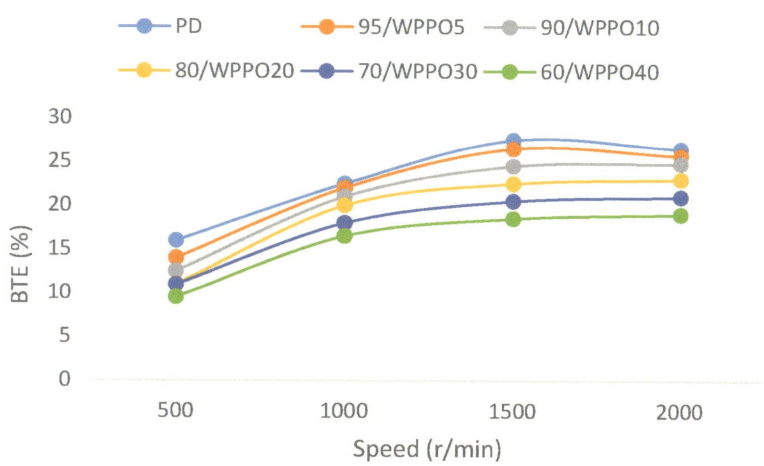

Figure 4.
Brake thermal efficiency versus speed.

The highest BTE value was 24. 5% by blend 95/WPPO5 at 1500 r/min (Mode 2, 75% engine load) compared to any other blend of WPPO. **Figure 4** shows values of 24.8%, 23%, 21% and 19% for full speed 2000 r/min 100% engine load, Mode 2) respectively for blends 90/WPPO10, 80/WPPO20, 70/WPPO30, and 60/WPPO40. However, blend 60/WPPO40/E25 reported the lowest values compared to the other blends. At 500 r/min (Mode 1, 25% engine load), the BTE value was 9.5% compared to 19% at full load 2000 r/min, Mode 2). These two are the lowest values of BTE for all the blends tested, as shown in **Figure 4**.

3.3 Carbon monoxide (CO)

Figure 5 is a variation of CO with two engine load modes (Mode 1, and Mode 2) with speed range of 500 r/min 25% engine load to 2000 r/min 100% engine load. The graph reveals that as the engine speed, load and the blend ratio increased, CO emissions reduced up to engine speeds of 1500 r/min (Mode 2, 75% of engine load). This was for PD and all blends 95/WPPO5, 90/WPPO10, 80/WPPO20, 70/WPPO30, and 60/WPPO40; the values were 270 ppm, 285 ppm, 315 ppm, 345 ppm, 370 ppm, 385 ppm respectively. The highest value of CO emission reported was 485 ppm for blend 60/WPPO40 and the lowest value reported was for blend 95/WPPO5 at 388 ppm.

Another observation is that as the engine was approaching full load (Mode 2, 2000 r/min), all the test fuels showed increased CO emissions with blends 95/WPPO5 and 90/WPPO10 reporting the lowest emissions value of 388 ppm and 435 ppm among the test blends across the entire engine load Modes 1 and 2 conditions. However, as the load increased from Mode 1 25% engine load speed of 500 r/min) to Mode 2 (75% engine load) the values reported were 320 ppm and 335 ppm respectively.

There are a number of factors, which explain the low CO emissions as the engine load and speed is increasing. The reason the blends show decreasing and increasing trends for Modes 1 and 2 respectively is due to high viscosity in WPPO. Viscosity affects the spray pattern resulting in poor fuel mixing therefore incomplete combustion and increased carbon monoxide emissions [59]. This phenomenon is linked to the increased engine load and the short ignition delay, hence increasing CO emissions. Additionally, the decrease in CO emissions could be due to the conversion of CO to CO_2 taking up this reaction from the high oxygen content of biodiesel [60].

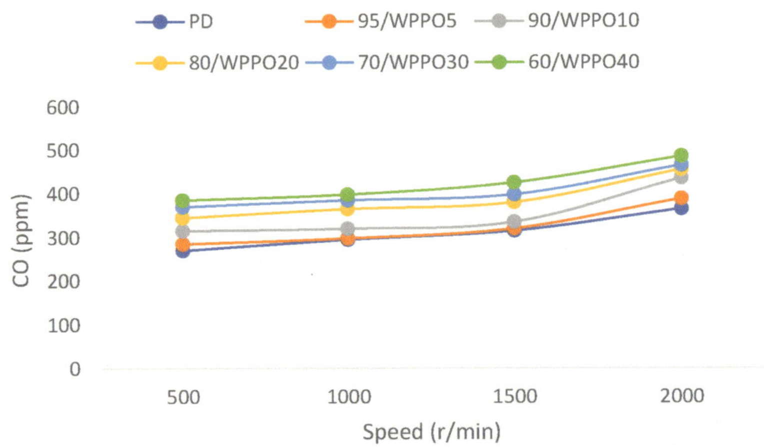

Figure 5.
Carbon monoxide versus speed.

3.4 Unburnt hydrocarbons (UHC)

UHC exhaust emissions are due to poor atomization after injection, over leaning zones and wall flame quenching [61, 62]. **Figure 6** is a variation of UHC emission with engine load. As the engine load and engine speed increased, the UHC emissions increased too. However, the increase was more significant as the engine load was in intermediate loads Mode 1 and Mode 2, 1000 r/min to 1500 r/min full load (50–75%). For example, at Mode 1, (1000 r/min, 50% engine load), the blend values were 23 ppm, 24 ppm, 26 ppm, 28 ppm, and 30 ppm respectively compared to full load Mode 2 2000 r/min) with 13 ppm, 15 ppm, 16 ppm, 18 ppm, and 26 ppm for blends 95/WPPO5, 90/WPPO10, 80/WPPO20, 70/WPPO30, and 60/WPPO40 respectively.

The blends 95/WPPO5 and 90/WPPO10 produced lower UHC emissions compared to the other test blends. The trends in **Figure 6** show high emission values for the blends compared to the PD test fuel values. However, the general trend in **Figure 6** shows that increased blend ratio significantly reduced UHC emissions across all the test fuels irrespective of the engine Mode. This reduction is due to the high oxygen of WPPO which has an oxygen content of 7.83 as shown in **Table 6** and in Section 3 of the results and discussion.

There are two main causes of increased hydrocarbon emissions, due to hydrogen radicals in diesel-WPPO blends and the presence of higher aromatic compounds [63]. Another contribution is caused by high density, low viscosity and low cetane of WPPO blends resulting in poor spray characteristics, leading to wall impingement, thus high UHC emissions. High blend ratio has also been identified as a factor that influences formation of UHC emissions using WPPO and has been reported by a number of researchers such as [64–66]. Hence the conclusion that high engine loads increases the values of UHC emissions proportionately to petroleum diesel. Increased UHC emissions can also be attributed to engine operating environment especially if the temperature range of 400–600°C exists in the combustion chamber. This is due to diesel exhaust pipe reaction, which either lowers or increases the concentration of UHC [67, 68].

3.5 Oxides of nitrogen (NO_X)

It is an established fact that NO_X emission are a function of in-cylinder temperature and atmospheric nitrogen, which is at 78% during intake. NOx emissions are also a

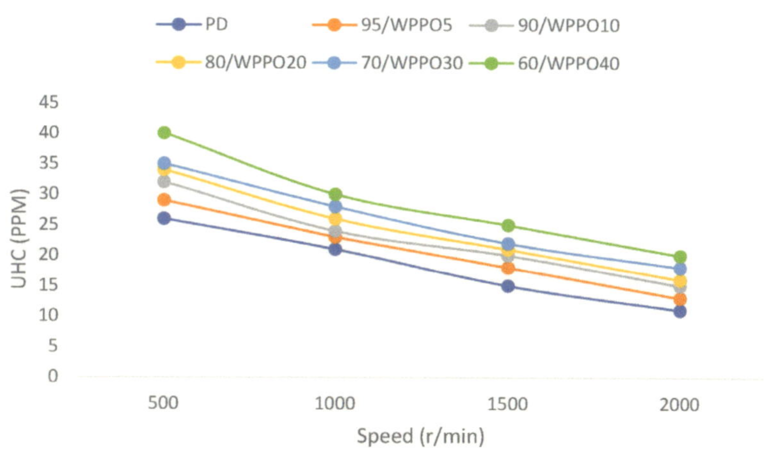

Figure 6.
Unburnt hydrocarbons versus speed.

function of three main mechanisms in the combustion theory [61, 69]. **Figure 7** is a variation of engine idling load with NO_X emissions. The figure shows that as the engine idling load was increased there was an increase in the NO_X emissions irrespective of fuel blend ratio. The values of NO_X emissions for the blends 95/WPPO5, 90/WPPO10, and 80/WPPO20 reported higher values at (Mode 2, 75% load) compared to Mode 1. For example, at 1500 the values of the blends were 335 ppm, 358 ppm, and 475 ppm, compared to PD fuel at 300 ppm.

Table 7 is showing different test fuel properties, units of measurement and testing standards used in this experiment. Blends 70/WPPO30 and 60/WPPO40 had the highest NO_X emissions compared to the other blends of 95/WPPO5, 90/WPPO10, and 80/WPPO20 across all the engine load conditions tested.

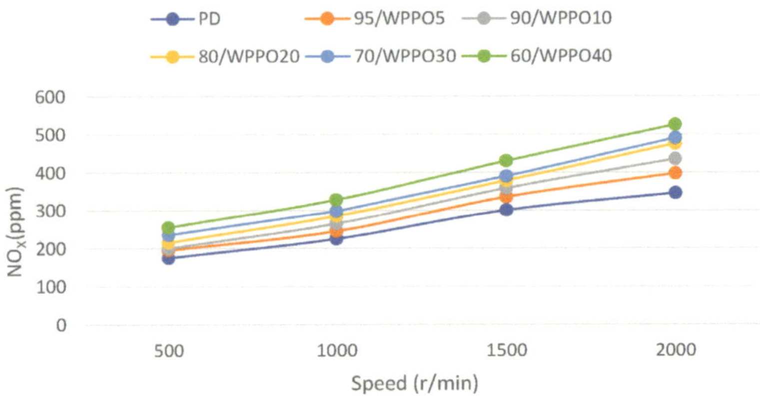

Figure 7.
Oxides of nitrogen emissions versus different engine speeds.

Property	Unit	PD	WPPO	Standard
Appearance	—	Clear/brown	Clear/amber	Visual
Density @ 20 °C	kg/M^3	838.8	788.9	ASTM D1298
Kinematic Visc @ 40 °C	mm^2/s	2.32	2.17	ASTM D445
Flash point	^{0}C	56.0	20.0	ASTM D93
Cetane index	—	46	65^a	ASTM D4737
Hydrogen	%	12.38	11.77	ASTM D7171
Cu corrosion	3 hrs @ 100°C	—	1B	ASTM D130
Carbon	%	74.99	79.60	ASTM D7662
Oxygen	%	12.45	7.83	ASTM D5622
Sulfur content	%	< 0.0124	0.15	ASTM D4294
IBP temperature	^{0}C	160	119	ASTM D86
FBP temperature	^{0}C	353.5	353.5	ASTM D86
Recovery	%		98	—
Residue and loss	%		2.0	—
Gross calorific value	MJ/kg	44.84	42.15^b	ASTM D4868

a and b are calculated values.

Table 7.
Test fuel biodiesel properties, units of measurement, testing standard methods and the values for PD compared to WPPO.

At 500 r/min 25% engine load (Mode 1), the two blends (70/WPPO30, 60/WPPO40) had values of 235 ppm and 255 ppm respectively. However, at full speed (2000 r/min, 100% engine load (Mode 2) the NO_X emissions for the two blends increased to 490 ppm and 525 ppm respectively compared to blends 95/WPPO5 for the same speed and engine load (500 r/min, Mode 1) at 175 ppm and at full load (2000 r/min, Mode 2) at 345 ppm.

As the blend ratio in **Figure 7** increased there was a direct increase in emissions of NO_X across all the blended fuels. However, blend 95/WPPO5 and 90/WPPO10 reported the lowest values of NO_X emissions (175 ppm and 195 ppm) compared to all the other tested blends experimented. The formation of NO_X in biodiesel fuel combustion strongly depends on the combustion temperatures and the oxygen concentration in the combustion zone. The high blend ratios of 80/WPPO20, 70/WPPO30, and 60/WPPO40 showed a shortened combustion process. Thus, a poor cooling effect and failure to decrease peak combustion temperatures leads to increased NO_X. WPPO blends emitted higher NO_X due to the higher cetane index compared to diesel fuel. High cetane index number fuels have a shorter ignition delay which means longer residence time at elevated chamber temperatures, hence higher NO_X compared to PD.

The increased NO_X emissions are a result of the presence of increased cetane index [70, 71] and other contaminants from the WPPO biodiesel impurities. Additionally, this could be due to the generation of radicals of hydrocarbons through molecular unsaturation in the blends, this being identical to the findings of [72, 73]. The final factor is due to increased chamber temperature which improves combustion but increases NO_X emissions, linked to the high oxygen content and the air fuel ratio factors [49].

4. WPPO combustion analysis

Due to the high cetane index of WPPO biodiesel the combustion process starts early compared to PD, hence a higher release rate than PD combustion. This leads to a higher cylinder peak pressure for WPPO biodiesel fuel compared to PD. Depicted in **Figure 8**, is a comparison of s WPPO blends with PD in Mode 1 25% engine load at speeds 500 r/min to 1000 r/min. Under this condition, WPPO blends in Mode 1

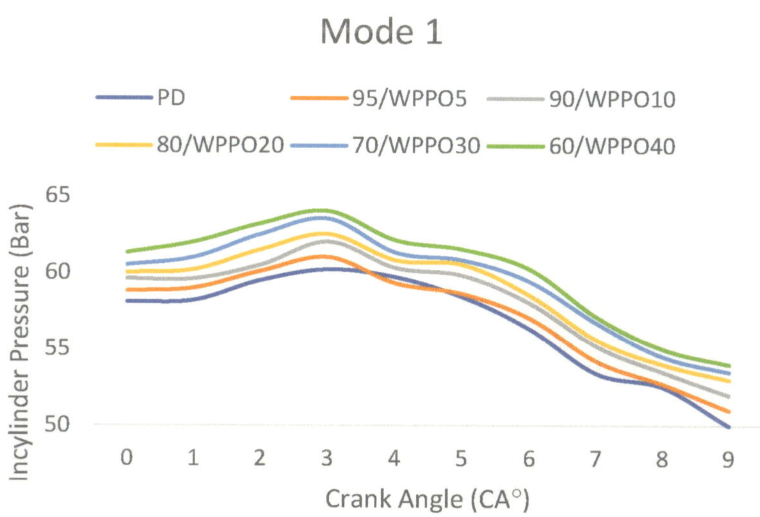

Figure 8.
In-cylinder pressure vs. crank angle variation compared to diesel and WPPO.

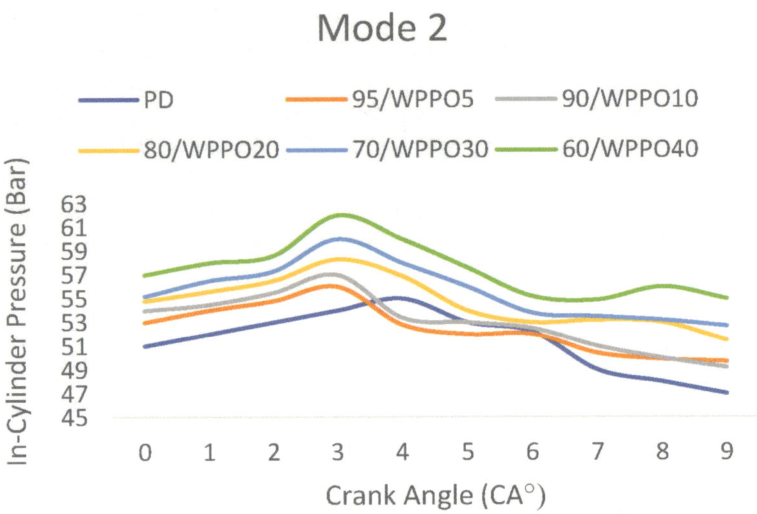

Figure 9.
In-cylinder pressure vs. crank angle variation compared to diesel and WPPO in mode 2.

exhibited higher peak cylinder pressure compared to PD, which is evident as the blend ratio increased as in **Figure 8**.

Compared to when the engine is running at high speed (high load), low speeds and low load residual gas temperatures and engine wall temperatures are low [69]. In other words, injection pressure and fuel temperature are low hence increased delay. This is the explanation why diesel in the combustion analysis starts after 3° CA compared to WPPO biodiesel blends. This causes diesel fuel to reach peak cylinder pressure after top dead center (TDC) in the power stroke. On the other hand, biodiesel blends reach peak cylinder pressure early, i.e., before TDC in the power stroke. For example, in **Figure 9** Mode 2 speed, the value for peak cylinder pressure for PD fuel is 55 bar compared to 56 bar for WPPO blend 95/WPPO5. This is due to enhanced combustion resulting from rapid combustion of the biodiesel blends at the pre-mixed phase. Of all the test fuels PD had the lowest peak cylinder pressure which occurs slightly after TDC [74].

5. Conclusion

- In the discussions in Section 3 under BSFC it was observed that, at high engine loads the conversion of heat energy to mechanical energy increased with increase in combustion temperature, leading to increased BSFC for the biodiesel. The increase was found to be proportional to the heating values of the different test fuels.

- Due to high densities exhibited by most biodiesel blends it was observed that blends suffer from high mass injection pressures which is in return increases the BSFCs of blends.

- As the percentage of the blend ratio increased there was a proportionate increase in engine fuel consumption due to lower calorific values of the blends. However, the values for all WPPO blends increased compared to PD test fuel.

- The Brake thermal efficiency of diesel engines was observed to be influenced by engine design, type of fuel used and the engine application. High engine load and speeds seems to increase BTE as can be seen in intermediate loads of 1000 r/min 50% engine load to 1500 r/min 75% engine load.

- Result obtained during this experiment show that the BTE increased as the load increased. This is explained by the reduction in the heat loss as the engine power (more fuel) increased with load.

- As the blend ratio and engine load and speed increased there was an increase in BTE across the blends of WPPO, but with a decrease in the BTE within the blends. This was attributed to the presence of aromatic compounds in waste pyrolysis plastic oil, which require a lot of energy to break.

- During experimentation it was observed that as the engine was approaching full load (Mode 2, 2000 r/min), all the test fuels showed increased CO emissions.

- As the engine load and engine speed increased, the UHC emissions increased too. However, the general trend in **Figure 6** shows that increased blend ratio significantly reduced UHC emissions across all the test fuels irrespective of the engine Mode.

- As the engine load and speed increased there was an increase in the NO_X emissions irrespective of fuel blend ratio.

- As the blend ratio increased there was a direct increase in emissions of NO_X across all the blended fuels.

Author details

Semakula Maroa* and Freddie Inambao
College of Agriculture Science and Engineering, Discipline of Mechanical Engineering, Green Energy Group, University of KwaZulu-Natal, Durban, South Africa

*Address all correspondence to: ssemakulamaroa@gmail.com

References

[1] Palash S, Kalam M, Masjuki H, Masum B, Fattah IR, Mofijur M. Impacts of Biodiesel Combustion on NOx Emissions and their Reduction Approaches. Renewable and Sustainable Energy Reviews 2013;23 473-490.

[2] Chisti Y. Biodiesel from Microalgae. Biotechnology Advances 2007;25(3) 294-306.

[3] Huang G, Chen F, Wei D, Zhang X, Chen G. Biodiesel Production by Microalgal Biotechnology. Applied Energy 2010;87(1) 38-46.

[4] Lee S, Yoshida K, Yoshikawa K. Application of Waste Plastic Pyrolysis Oil in a Direct Injection Diesel Engine: for a Small-Scale Non-Grid Electrification. Energy and Environment Research 2015;5(1) 18.

[5] Mani M, Nagarajan G, Sampath S. Characterization and Effect of Using Waste Plastic Oil and Diesel Fuel Blends in Compression Ignition Engine. Energy 2011;36(1) 212-219.

[6] Maroa S, Inambao F. Influence of Exhaust Gas Recirculation and Emission Characteristics of a Diesel Engine Using Pyrolyzed Waste Plastic Biodiesel and Blends. International Journal of Applied Engineering Research 2018;13(10) 8321-8335.

[7] Phan AN, Phan TM. Biodiesel Production from Waste Cooking Oils. Fuel 2008;87(17-18) 3490-3496.

[8] Rahman SA, Masjuk, H, Kalam M, Abedin M, Sanjid A, Imtenan S. Effect of Idling on Fuel Consumption and Emissions of a Diesel Engine Fueled by Jatropha Biodiesel Blends. Journal of Cleaner Production 2014;69 208-215.

[9] Mofijur M, Masjuki HH, Kalam MA, Atabani AE, Shahabuddin M, Palash SM, Hazrat MA. Effect of Biodiesel from Various Feedstocks on Combustion Characteristics, Engine Durability and Materials Compatibility: A review. Renewable and Sustainable Energy Reviews 2013;28 441-455.

[10] Mofijur M, Masjuki HH, Kalam M, Hazrat MA, Liaquat AM, Shahabuddin M, Varman M. Prospects of Biodiesel from Jatropha in Malaysia. Renewable and Sustainable Energy Reviews 2012;16(7) 5007-5020.

[11] Rahman SA, Masjuk, H, Kalam M, Abedin M, Sanjid A, Rahman MM. Assessing Idling Effects on a Compression Ignition Engine Fueled with Jatropha and Palm Biodiesel Blends. Renewable Energy 2014;68 644-650.

[12] Eurostatistics. Environmental Tax Statistics, European Commission Eurostat; 2018. http://epp.eurostat.ec.europa.eu/statistics_explained/index.php/Environmental_tax_statistics (accessed 09 September 2019).

[13] Atabani A, Silitonga A, Ong H, Mahlia T, Masjuki H, Badruddin IA, Fayaz H. Non-Edible Vegetable Oils: A Critical Evaluation of Oil Extraction, Fatty Acid Compositions, Biodiesel Production, Characteristics, Engine Performance and Emissions Production. Renewable and Sustainable Energy Reviews 2013;18 211-245.

[14] Fazal M, Haseeb A, Masjuki H. Biodiesel Feasibility Study: An Evaluation of Material Compatibility; Performance; Emission and Engine Durability. Renewable and Sustainable Energy Reviews 2011;15(2) 1314-1324.

[15] Hussan MJ, Hassan MH, Kalam MA, Memon LA. Tailoring Key Fuel Properties of Diesel–Biodiesel–Ethanol Blends for Diesel Engine. Journal of Cleaner Production 2013;51 118-125.

[16] Xue J, Grift TE, Hansen A. C. 2011 Effect of Biodiesel on Engine Performances and Emissions. Renewable and Sustainable Energy Reviews 2011; 15(2) 1098-1116.

[17] Ananthakumar S, Jayabal S, Thirumal P. Investigation on Performance, Emission and Combustion Characteristics of Variable Compression Engine Fuelled with Diesel, Waste Plastics Oil Blends. Journal of the Brazilian Society of Mechanical Sciences and Engineering 2017;39(1) 19-28.

[18] Kaimal VK, Vijayabalan P. A Detailed Study of Combustion Characteristics of a DI Diesel Engine Using Waste Plastic Oil and its Blends. Energy conversion and Management 2015;105 951-956.

[19] Naima K, Liazid A. Waste Oils as Alternative Fuel for Diesel Engine: A Review. Journal of Petroleum technology and Alternative fuels 2013;4(3) 30-43.

[20] Rahman SA, Masjuki H, Kalam M, Abedin M, Sanjid A, Sajjad, H. Production of Palm and *Calophyllum inophyllum* Based Biodiesel and Investigation of Blend Performance and Exhaust Emission in an Unmodified Diesel Engine at High Idling Conditions. Energy Conversion and Management 2013;76 362-367.

[21] Roy MM, Wang W, Bujold J. Biodiesel Production and Comparison of Emissions of a DI Diesel Engine Fueled by Biodiesel–Diesel and Canola Oil–Diesel Blends at High Idling Operations. Applied Energy 2013;106 198-208.

[22] Calcagno JA. Evaluation of Heavy-Duty Diesel Vehicle Emissions During Cold-Start and Steady-State Idling Conditions and Reduction of Emissions from a Truck-Stop Electrification Program, Proceedings of the Air and Waste Management Association's Annual Conference and Exhibition, AWMA, 2005.

[23] Lambert DC, Vojtisek-Lom M, Wilson PJ. 2002 Roadside Emissions Study–Preliminary Results for Stationary and On-Road Testing of Diesel Trucks in Tulare, CA. Clean Air Technologies International, Inc. in cooperation with California Air Resource Board, Mobile Source Operations Division.

[24] Khan AS, Clark NN, Thompson GJ, Wayne WS, Gautam M, Lyon DW, Hawelti D. Idle emissions from Heavy-Duty Diesel Vehicles: Review and Recent Data. Journal of the Air & Waste Management Association 2006;56(10) 1404-1419.

[25] Roy MM, Calder J, Wang W, Mangad A, Diniz, FCM. Cold Start Idle Emissions from a Modern Tier-4 Turbo-Charged Diesel Engine Fueled with Diesel-Biodiesel, Diesel-Biodiesel-Ethanol, and Diesel-Biodiesel-Diethyl Ether Blends. Applied Energy 2016;180 52-65.

[26] Zietsman J, Villa JC, Forrest TL, Storey JM. 2005 Mexican Truck Idling Emissions at the El Paso-Ciudad Juarez Border Location. https://tti.tamu.edu/tti-publication/mexican-truck-idling-emissions-at-the-el-paso-ciudad-juarez-border-location/ (accessed 9 September 2019).

[27] Chaichan MT. Experimental Study of the Effect of Exhaust Gas Recirculation (EGR) and Injection Timing on Emitted Emissions at Idle Period. Al-Khwarizmi Engineering Journal 2014;10(4) 33-44.

[28] Chin J-Y, Batterman SA, Northrop WF, Bohac SV, Assanis, DN. Gaseous and Particulate Emissions from Diesel Engines at Idle and Under Load: Comparison of Biodiesel Blend and Ultralow Sulfur Diesel Fuels. Energy & Fuels 2012;26(11) 6737-6748.

[29] Shancita I, Masjuki H, Kalam M, Fattah IR, Rashed M, Rashedul H. A Review on Idling Reduction Strategies to Improve Fuel Economy and Reduce Exhaust Emissions of Transport Vehicles. Energy Conversion and Management 2014;88 794-807.

[30] Storey JME., Thomas JF, Lewis SA, Dam TQ, Edward, KD, DeVault GL, Retrossa DJ. Particulate Matter and Aldehyde Emissions from Idling Heavy-Duty Diesel Trucks. SAE Paper 2003-01-0289; 2003.

[31] Brodrick C-J, Dwyer HA, Farshch, M, Harris DB, King Jr FG. Effects of Engine Speed and Accessory Load on Idling Emissions from Heavy-Duty Diesel Truck Engines. Journal of the Air & Waste Management Association 2002;52(9) 1026-1031.

[32] Denning C, Camille K. The Good Haul: Innovations that Improve Freight Transportation and Protect the Environment; 2010. https://trid.trb.org/view/914909 (accessed 09 September 2020.

[33] Gaines L, Vyas A, Anderson JL. 2006 Estimation of Fuel use by Idling Commercial Trucks. Transportation Research Record 2006;1983(1) 91-98.

[34] Carrico AR, Padgett P, Vandenbergh MP, Gilligan J, Wallston, KA. Costly Myths: An Analysis of Idling Beliefs and Behavior in Personal Motor Vehicles. Energy Policy 2009;37(8) 2881-2888.

[35] Nitschke B, Wagner F, Anderson M, Slezak L. Idle Reduction Technology Demonstration Plan (No. NREL/TP-540-33629) Retrieved from Golden, Colorado, United States; 2003.

[36] Jayed M, Masjuki HH, Saidur R, Kalam M, Jahirul MI. Environmental Aspects and Challenges of Oilseed Produced Biodiesel in Southeast Asia. Renewable and Sustainable Energy Reviews 2009;13(9) 2452-2462.

[37] Liaquat A, Kalam M, Masjuk, H, Jayed M. Potential Emissions Reduction in Road Transport Sector Using Biofuel in Developing Countries. Atmospheric Environment 2010;44(32) 3869-3877.

[38] Candeia R, Silva M, Carvalho Filho J, Brasilino M, Bicudo T, Santos I, Souza A. Influence of Soybean Biodiesel Content on Basic Properties of Biodiesel–Diesel Blends. Fuel 2009;88(4) 738-743.

[39] Demirbas A. Biodiesel: Cham: Springer; 2008.

[40] Singh SP, Singh D. 2010 Biodiesel Production through the use of Different Sources and Characterization of Oils and their Esters as the Substitute of Diesel: A Review. Renewable and Sustainable Energy Reviews 2010 14(1) 200-216.

[41] Jain S, Sharma MP. Oxidation Stability of Blends of Jatropha Biodiesel with Diesel. Fuel 2011;90(10) 3014-3020.

[42] Maroa S, Inambao F. The Effect of Cetane Number and Oxygen Content in the Performance and Emissions Characteristics of a Diesel Engine Using Biodiesel Blends. Journal of Energy in Southern Africa 2019;30(2) 1-13.

[43] Lapuerta M, Armas O, Rodriguez-Fernandez J. Effect of Biodiesel Fuels on Diesel Engine Emissions. Progress in Energy and Combustion Science 2008;34(2) 198-223.

[44] Lapuerta M, Rodríguez-Fernández J, Agudelo J. R. Diesel Particulate Emissions from used Cooking Oil Biodiesel. Bioresource Technology 2008;99(4) 731-740.

[45] Bhale PV, Deshpande NV, Thombre SB. Improving the Low Temperature Properties of Biodiesel Fuel. Renewable Energy 2009;34(3) 794-800.

[46] Cetinkaya M, Ulusoy Y, Tekìn Y, Karaosmanoğlu F. Engine and Winter Road Test Performances of Used Cooking Oil Originated Biodiesel. Energy Conversion and Management, 2005;46(7-8) 1279-1291.

[47] Maroa, S., Inambao F. The Production of Pyrolysis Oil from Waste Plastic and Other Solid Waste Biomass Design, Fabrication and Evaluation. Paper presented at the African And International Domestic Use of Energy Conference, Cape Town; 2019.

[48] Azam MM, Waris A, Nahar N. Prospects and Potential of Fatty Acid Methyl Esters of Some Non-Traditional Seed Oils for use as Biodiesel in India. Biomass and Bioenergy 2005;29(4) 293-302.

[49] Devan P, Mahalakshmi N. A Study of the Performance, Emission and Combustion Characteristics of a Compression Ignition Engine Using Methyl Ester of Paradise Oil–Eucalyptus Oil Blends. Applied Energy 2009;86(5) 675-680.

[50] Lapuerta M, Armas O, Hernández JJ, Tsolakis A. Potential for Reducing Emissions in a Diesel Engine by Fuelling with Conventional Biodiesel and Fischer–Tropsch Diesel. Fuel 2010;89(10) 3106-3113.

[51] Raheman H, Ghadge SV. Performance of Compression Ignition Engine with Mahua (*Madhuca indica*) Biodiesel. Fuel 2007;86(16) 2568-2573.

[52] Tüccar G, Tosun E, Özgür T, Aydın K. 2014 Diesel Engine Emissions and Performance from Blends of *Citrus sinensis* Biodiesel and Diesel Fuel. Fuel 2014;132 7-11.

[53] Armas O, Yehliu K, Boehman AL. Effect of Alternative Fuels on Exhaust Emissions During Diesel Engine Operation with Matched Combustion Phasing. Fuel 2010;89(2) 438-456.

[54] Godiganur S, Murthy CS, Reddy RP. Performance and Emission Characteristics of a Kirloskar HA394 Diesel Engine Operated on Fish Oil Methyl Esters. Renewable Energy 2010;5(2) 355-359.

[55] Zhu L, Zhang W, Liu W, Huang Z. 2010 Experimental Study on Particulate and NOx Emissions of a Diesel Engine Fueled with Ultra Low Sulfur Diesel, RME-Diesel Blends and PME-Diesel Blends. Science of the Total Environment 2010; 408(5) 1050-1058.

[56] Ramalingam S, Rajendran S. 2019 Assessment of Performance, Combustion, and Emission Behavior of Novel Annona Biodiesel-Operated Diesel Engine. In: Azad K (ed.) Advances in Eco-Fuels for a Sustainable Environment. Amsterdam: Elsevier; 2019. p391-405.

[57] Kalargaris I, Tian G, Gu S. Combustion, Performance and Emission Analysis of a DI Diesel Engine Using Plastic Pyrolysis Oil. Fuel Processing Technology 2017;157 108-115.

[58] Uddin SA, Azad AK, Alam M, Ahamed J. Performance of a Diesel Engine Run with Mustard-Kerosene Blends. Procedia Engineering 2015;105 698-704.

[59] Ghurri A, Kim J-D, Kim HG, Jung, J-Y, Song K-K. The Effect of Injection Pressure and Fuel Viscosity on the Spray Characteristics of Biodiesel Blends Injected into an Atmospheric Chamber. Journal of Mechanical Science and Technology 2012;26(9) 2941-2947.

[60] Rahman MM, Hassan MH, Kalam MA, Atabani AE, Memon LA, Rahman, SA. Performance and Emission Analysis of *Jatropha curcas* and *Moringa oleifera* Methyl Ester Fuel Blends in a Multi-Cylinder Diesel Engine. Journal of Cleaner Production 2014;65 304-310.

[61] Heywood JB. 2012 Internal Combustion Engine Fundamentals

(Vol. 17). New Delhi, India: McGraw Hill Education India.

[62] Mendez S, Kashdan JT, Bruneaux G, ThirouardB,VangraefschepeF.Formation of Unburned Hydrocarbons in Low Temperature Diesel Combustion. SAE International Journal of Engines 2010;2(2) 205-225.

[63] Murugan S, Ramaswamy M, Nagarajan G. The use of Tyre Pyrolysis Oil in Diesel Engines. Waste Management 2008;28(12) 2743-2749.

[64] Lujaji, F, Kristóf L, Bereczky A, Mbarawa M. Experimental Investigation of Fuel Properties, Engine Performance, Combustion and Emissions of Blends Containing Croton Oil, Butanol, and Diesel on a CI Engine. Fuel 2011;90(2) 505-510.

[65] Maroa S, Inambao F. Effects of Biodiesel Blends Varied by Cetane Numbers and Oxygen Contents on Stationary Diesel Engine Performance and Exhaust Emissions. In: Wos P, Jakubowski M (eds) Numerical and Experimental Studies on Combustion Engines and Vehicles. Rijeka: IntechOpen; 2020. DOI: 10.5772/intechopen.92569

[66] Tutak W, Lukacs K, Szwaja S, Bereczky, A. Alcohol–Diesel Fuel Combustion in the Compression Ignition Engine. Fuel 2015;154 196-206.

[67] Sanli H, Canakci M, Alptekin E, Turkcan A, Ozsezen AN. Effects of Waste Frying Oil Based Methyl and Ethyl Ester Biodiesel Fuels on the Performance, Combustion and Emission Characteristics of a DI Diesel Engine. Fuel 2015;159 179-187.

[68] Shirneshan A. HC, CO, CO2 and NOx Emission Evaluation of a Diesel Engine Fueled with Waste Frying Oil Methyl Ester. Procedia-Social and Behavioral Sciences 2013;75 292-297.

[69] Turns SR. An Introduction to Combustion: Concepts and Applications (3 ed.) New Delhi, India: McGraw-Hill; 1996.

[70] Hoekman SK, Broch A, Robbins C, Ceniceros E, Natarajan M. Review of Biodiesel Composition, Properties, and Specifications. Renewable and Sustainable Energy Reviews 2012;16(1) 143-169.

[71] Robbins C, Hoekman SK, Gertler A, Broch A, Natarajan M. 2009 Biodistillate Transportation Fuels 2.-Emissions Impacts (No. 2009-01- 2724). SAE Technical Paper.

[72] Altun Ş. Effect of the Degree of Unsaturation of Biodiesel Fuels on the Exhaust Emissions of a Diesel Power Generator. Fuel 2014;117 450-457.

[73] Benjumea P, Agudelo JR, Agudelo AF. Effect of the Degree of Unsaturation of Biodiesel Fuels on Engine Performance, Combustion Characteristics, and Emissions. Energy & Fuels 2010;25(1) 77-85.

[74] Tarabet L, Loubar K, Lounici M, Hanchi S, Tazerou, M. Experimental evaluation of Performance and Emissions of DI Diesel Engine Fuelled with Eucalyptus Biodiesel. Proceedings of the Internal Combustion Engines: Performance, Fuel Economy and Emissions 2011;167-176.

Chapter 6

Mitigation of Emissions through Injection Strategies for C I Engine

Jayashri N. Nair

Abstract

Fuel conversion efficiency is high with diesel engines compared to petrol engines. However high emissions from diesel is a matter of concern and its mitigation paves way for scope of research. Exhaust gas recirculation is one of the method widely accepted to curb NO_x emissions. Recently, split or multiple-injection strategy has been explored by researchers to precisely control the fuel injected per cycle and also to mitigate emissions. Present work reflects technical review of effect of injection strategies on performance, emissions and combustion on C.I. engine with diesel and biodiesel as fuel. Injection strategies like duration of injection, number of injections, the dwell period between two injections, quantity of injection, and multiple injections are analyzed for their influence on engine output and brake specific fuel consumption. Also their effect on emissions especially soot and NO_x emission are reviewed. First the effect of injection strategies with diesel fuel is discussed followed by biodiesel.

Keywords: split injection, pilot injection, NO_x emissions, Soot emissions, multiple injections

1. Introduction

The emissions from diesel engine are hazardous for human health as it causes respiratory problem in human beings. NO_x emissions in combination with water and oxygen forms acid rain and is also responsible for the global warming. To improve the air quality in cities, regulations and control measures are taken to lower exhaust emissions from heavy vehicles. However, there are no much improvements mostly because of the difficulties in eliminating NO_x and PM emissions. The need of an hour is simultaneous reduction of NO_x and PM emissions from diesel engines. However it is challenging to control NO_x and PM emission simultaneously. Optimizing injection timing through split injection can be a prominent solution to curb these emissions from engine. In split injection, the injection is divided into two or more pulses thus reducing the delay time. Major fraction of the combustion takes place during expansion stroke. NO_x released from the engine is reduced as NO_x is mostly formed in the premixed combustion. Investigations are done by many researchers to optimize the injection scheme and analyze the effect of split injections on combustion process and emissions.

Fuel injection timings influence the performance, emissions and the combustion part of the engine to a great extent. With the variation in the injection timings the state of air changes which effects the ignition delay. Occurrence of injection when the piston is far away from TDC reduces the temperature and pressure of the air

IntechOpen

increasing the ignition delay. If injection starts when piston is nearer to TDC then temperature and pressure of the air will be higher which will decrease the ignition delay.

2. Multiple injections

Multiple fuel injections are employed, instead of single injection as in the conventional fuel injection system. The main injection provides maximum quantity of fuel per engine cycle. Pre injections can be one or more injections before the main injection. In many cases pre injections are termed as pilot injection. Pre injections can be employed when the compression stroke starts or can also be employed just before the main injection.

Multiple injections strategies involve following phenomena

1. Introducing fuel and dividing heat release rate (HRR).

2. Spatial distribution of fuel in combustion chamber.

3. Cooling effect of vaporizing fuel.

Linearity of the heat release rate is desirable for minimizing combustion noise. Pre injections helps in dividing the heat release process which eventually controls the combustion noise. Triple and quadruple injections also assists in minimizing combustion noise. Furthermore spatial fuel distribution is improved due to multiple injection strategies for effective usage of air for combustion in combustion chamber. Generally, this effect can lead to a reduction in particulate emissions at intermediate engine loads. The cooling effect associated with fuel vaporization lowers locally and globally the temperature of the gases contained in the combustion chamber. These phenomena can be applied to modify the rate of heat release in the early period of combustion by increasing the ignition there by allowing for a longer duration of mixing period and thus a more homogeneous fuel/air mixture. Correct use of cooling effect provides a valuable degree of freedom in the optimization of the noise, emission and fuel economy trade off. Multiple injections can be used for lower compression ratio engines with high EGR rates to mitigate the noise, emission and fuel economy trade off.

Post injections are injections after the main injection. It can occur immediately after the main injection or long time after the main injection and are also referred as after injections or late injections. Post injections are used to mitigate unburned hydrocarbons, particulate matter and for managing exhaust after treatment. Post injections can also be used with various EGR rates with single or double post injections to improve trade off relationship between NO_x and HC, CO emissions.

3. Effect of multiple injections on CI engine with diesel fuel

3.1 Effect of varying injection timing and multiple injections with diesel fuel

In the fullness of time, clean, cheap and high torque diesel engines will be in demand. Controlled combustion is an essential part for an economic and clean diesel engine. Homogeneous air- fuel mixture plays a key role in effective combustion. Homogeneity of air-fuel mixture is affected by spatial fuel distribution in combustion chamber. Higher injection pressure, longer dwell time and increased pilot fuel quantities can contribute to better mixing quality resulting in increased homogeneity of air-fuel mixture and optimum engine performance with low fuel consumption and soot emissions. However trade off does exists between NO_x and soot emissions with more quantity of fuel in pilot injection. With changing quantity of fuel in first injection, reduction in both particulate emissions and BSFC was observed with single injection at fast rate [1]. Also optimizing the rate of injection for the lowest BSFC did not affect the pressure rise with single injection. However, it was evident that the quantity of fuel in the first injection effected the rate of in-cylinder pressure rise and NO_x emissions also. Peak pressure reduced by more than 45% with split injection. The NO_x emissions increased with more quantity of fuel in the first injection, whereas the particulate emissions decreased. High homogeneity of air fuel mixture leads to early burning of fuel in premixed combustion zone, increasing the in- cylinder combustion temperature resulting in high NO_x emissions. Furthermore, split injection allows combustion to continue in the power stroke without any increase in soot emissions and thus utilizing the maximum charge. This indicates that pulsed injection may provide a means to reduce particulate emissions, and allow for reduced NO_x from controlled pressure rise.

In another investigation, the injection pressure and the total quantity of fuel injected was kept constant but the ratio of amounts of fuel injected between two injection pulses were varied and also the time intervals between the two injections were varied [2]. With longer time interval between two injections and less quantity of fuel in the first injection, the in-line cylinder pressure decreased. Mixing of in-cylinder gas was enhanced by the second injection and the injection interval, which eliminated local high temperature areas resulting in reduced NO_x emissions. Shorter time interval between two injections resulted in increased NO_x emissions. The smoke emissions were high with less fuel quantity in the first injection due to the fact that large amount of oxygen was consumed during the first injection. More fuel was injected in the second injection which led to decreased combustion duration and resulted in increased smoke. However the smoke emission reduced with increased fuel quantity in the first injection. BSFC improved with less fuel quantity in second injection and shorter injection interval resulting in delayed crank angle at 50% heat release.

Slight modification in the engine can assist in optimizing and treating the tradeoff between NO_x and other emissions in combination with split injections. Design of double lobe cam with optimum split and dwell can influence reduction in engine emissions [3]. Use of double lobed cam showed reduction in NO_x emission but showed reverse trend on CO and CO_2 emissions. Split injection with 10° crank angle dwell using double lobed cam was optimum in effectively reducing the NO_x emission by 14.1% and unburned hydro carbon by 11.8% without penalizing the soot and power. There was no considerable variation observed with SFC and brake thermal efficiency for the modified engine with split injection. In a similar study the amount of fuel injected was divided equally for pilot injection and main injection and was compared with the single injection [4]. Injection duration and injection rates were reduced. In case of split injection it was observed that the heat release rate reduced compared to the single injections resulting in lower NO_x emissions. On retarding the split injection timing, soot emissions decreased compared to single injection. As the split injection timing was retarded, the larger size particles which forms the total particle volume and weight decreased, but HC and CO emissions were higher for split injections.

Factors influencing emissions namely, ignition delay, adhered fuel, and squish are affected with the changes in the injection timings. Changes in injection timings affects the position of the piston, cylinder pressure and temperature at injection. Advancement of first stage injection timing to 80 °C BTDC, reduced BSFC to 20% [5].

The NO_x and smoke emissions were improved with less quantity of fuel in the first injection and advancement of the injection from 80 °C BTDC to 100 °C BTDC. The CO emissions decreased when the first injection was advanced to 100 °C BTDC and the second injection was retarded over TDC.

3.2 Effect of varying fuel injection pressure with diesel fuel

Increasing fuel injection pressure is one of the best control techniques for optimizing combustion in diesel engines. High injection pressure promotes fine atomization of fuel, and uniform mixing of fuel and air thus, decreasing the combustion duration. Split injections coupled with higher pressure injection strategy tend to improve performance and emissions of engine. However combination of this strategy results in high NO_x emissions. Also increased spray tip penetration due to high injection pressure results in spray impingement on wall and piston, leading to high HC emissions. In an investigation, conventional diesel fuel (Navy NATO F76) was compared with the new Navy hydro processed renewable diesel (HRD) fuel from algal sources, as well as the high cetane reference fuel nC16 (n-hexadecane CN¼100) [6]. It was observed that increasing fuel injection pressure shortened the ignition delay for all fuels. The combustion duration was longer for higher cetane number fuels for the same fuel pressure. This may be attributed to less premixing of fuel before the start of combustion. The author tried to study the relation between physical and chemical delay times with HRD and nC16. As injection pressure increases, the importance of chemical delay increases especially with lower cetane number. Split injections combined with exhaust gas recirculation (EGR) and higher injection pressure has the potential for diminishing NO_x emissions and elevating the engine performance. The test was carried out on a V6 common rail direct injection engine [7]. The engine was turbocharged with variable turbine geometry turbochargers. The experiments were carried out at two different speeds 1500 rpm (35.1, 70.2 and 140 Nm) and 2000 rpm (43.3, 86.6 and 120 Nm) with injection pressure of 300, 430, 500, 600, 700 bar. The increase in injection pressure from 300 bar to 700 bar resulted in improved engine performance and emissions for all engine conditions. Increase in peak cylinder was observed for both EGR ON and EGR off conditions. In all engine test conditions the BSFC decreases as the injection pressure increases due to an improved fuel mixture and rapid combustion rate. However, these values were slightly increased when the engine operates with cooled EGR ON (1500 rpm, 35.1 Nm EGR OFF) at 600–700 bar injection pressure. This is strongly believed to be due to decreased combustion temperature and reduced oxygen availability for EGR ON. This results in higher fuel consumption. The reduced oxygen, higher inert in-cylinder gas and low boost pressure contributed to the poor combustion with cooled EGR ON. The specific fuel consumptions also decreased at both speeds. Reduction in CO emissions were observed with EGR off by 80% at 1500 rpm and 60% at 2000 rpm. The THC emissions also showed reduction with EGR OFF by approximately 70% at 1500 rpm and approximately 90% at 2000 rpm. However the NO_x emissions increased 4 times at 1500 rpm especially at 70.2 Nm and by approximately 3 times at 2000 rpm especially at 43.4 Nm. The rate that NO_x emissions increased as injection pressure rise was lower for EGR ON than for EGR OFF. Several factors that contribute to the fast combustion process in diesel engines are; fuel droplet size, penetration length, turbulence intensity, fuel evaporation, rate of combustion and ignition delay. However, all of these can contribute to the higher NO_x formation. As expected, the EGR ON produces improved NO_x reduction even at higher injection pressure for all engine test conditions. The use of cooled EGR produces lower peak in-cylinder pressure compared to cooled EGR OFF. This is due to a retarded combustion, a result of the reduced oxygen density and high heat capacity of the gas mixture. In addition, the reduction of fuel burnt in the premixed combustion phase can be considered as a strong factor in lowering in-cylinder peak pressure.

4. Effect of multiple injections on CI engine with biodiesel fuel

4.1 Effect of varying injection timing and multiple injections with biodiesel fuel

The formation of NO_x with biodiesel fuel blends, depend upon bulk modulus of biodiesel, oxygen content of biodiesel, cetane number, saturated fatty acid content and engine operating conditions. Varying injection timing has proved to be a potential technique in curbing NO_x emissions in case of biodiesel. During the premixed combustion phase, NO_x is formed in lean flame region. Optimization of combustion timing and regulating the quantity of fuel burned in premixed phase can significantly reduce the NO_x emissions.

Coconut biodiesel with 20% and 50% proportions by volume in diesel were tested with single, double and triple injections [8]. In double injection the fuel mass was distributed equally for both injections while in triple injection 33.33% of mass was distributed in each injections with a dwell time of 1.3 ms in both double and triple injection case. Injection timing was varied from 12°BTDC to 2°ATDC. Advanced start of injection timings increased brake thermal efficiency for all fuels. This may be due to longer ignition delay which promotes better mixing of fuel and air resulting in efficient combustion. Also in case of advanced injections the peak pressure was attained at TDC. With single injection at 12°BTDC brake thermal efficiency decreased with the increase in the biodiesel blends. For all injection timings and schemes the BSFC was higher for biodiesel blends in comparison to diesel. Due to the lower heating value of the biodiesel more mass is needed to produce the same power output. Also it is noted that with increase in split injections the BSFC increases for all the fuels. This may be due to the longer duration of combustion process with the higher number of split injections which increases the heat loss and reduces the peak pressure which reduces the work output. NO_x is increased with increase in the advancement of injections due to occurrence of peak pressure around TDC which results in higher temperature. Biodiesel blends showed less NO_x emissions than diesel for all split injections and single injection. This may be due to higher cetane number and lower calorific values of the blends which helps in reducing the rate of heat release. With B50 blend simultaneous reduction of NO_x and smoke emission is possible with retardation in injection timing with triple injection scheme.

In another investigation effect of injection timing on NO_x emissions with ultra-low sulfur diesel fuel and biodiesel blends (name not specified) were studied [9]. At low load conditions biodiesel blends were found to produce slightly lower NO_x emissions than the neat diesel fuel. However at higher loads the NO_x emissions were higher for both single and double injection conditions. Overall, biodiesel diesel blends and the baseline diesel fuel had very similar heat release rate profiles. When the injection timing was retarded it proved effective for single injection conditions than using pilot injection with retarded main injection in reducing the NO_x emissions. Under the low load condition, the pilot injection strategy led to substantially reduced NO_x emissions. Similar effect of retardation on NO_x emission is reported using Pongmia methyl ester [10]. The injection timing was retarded by 4° for the blends. All the blends (B20, B40, B60, and B80) showed lower NO_x emissions and HC emissions with retarded injection timing. The smoke emissions also reduced for all the blends with retarded injection timing.

4.2 Effect of varying quantity of fuel in the injections with biodiesel fuel

Varying quantity of fuel in multiple injections affects the performance and emission of the engine [11]. For single injection the quantity of fuel discharged was 10 mg whereas for multiple injections the fuel quantity was divided in equal proportion for both injections in first test (5 mg + 5 mg) and 3 mg in first injection and 7 mg in second injection for second test. With single injection start of energizing was arranged at TDC, 10°BTDC, 20°BTDC whereas for multiple injections it was arranged at TDC, 10°BTDC, 20°BTDC, 30 ° BTDC. In case of single injection test, injection pressures varied to 60 MPa and 120 MPa. For single injection as the advancement increased the peak combustion pressure and peak heat release rate increased. However at 30 ° BTDC these values dropped. This can be attributed to the inline cylinder pressure and temperature which were lower during the spray and resulted in long ignition delay. Also the fuel injected dispersed in the squish and crevice regions. Soot emissions were lower with higher injection pressure. Peak NO_x emission was reported at 20°BTDC for 60 MPa and 15°BTDC for 120 MPa. For injection timings below these values, the NO_x emissions continuously decreased. The increase in NO_x emission was attributed to the proper mixing of the spray with air. The decrease in the NO_x emission at less injection timings was attributed to the incomplete combustion. CO and HC emissions increased with the advancement of the injection timings. In case of multiple injection at 120 MPa with 20°BTDC pilot injection and main injection at TDC showed lower heat release rates than single injection. Due to shorter ignition delay the second combustion resumed quickly than the first combustion consequently the temperature of combustion chamber increased for first combustion. During the first injection the consumption of oxygen will be maximum and so the rate of heat release was higher for the first injection as compared to second injection. It was observed that with shorter ignition timings supported less soot, HC and CO emissions but NO_x increased.

4.3 Effect of varying fuel injection pressure with biodiesel fuel

Proportion of biodiesel in diesel influences the injection spray pattern [12]. Due to the increased fuel line pressure, with increase of biodiesel in the blends the penetration distance of the spray also increases. Simulation results reports that the probability of wall impingement is more with higher blends. The ignition delay period decreased with all biodiesel blends resulting in less rate of pressure rise. This may attributed to higher cetane number of the biodiesel. Reduction in torque is noted with B100 at rated load. However for rest of the blends there was no significant change in the torque. HC, CO and smoke emissions decreased with all blends while NO_x emissions increased in the range of 1.4–22.8% with all biodiesel. This may be attributed to the oxygenated fuel and automatic advance in dynamic injection timing. It was concluded that B15 was the optimum blend with respect to no wall impingement and NO_x emissions.

Tests were conducted to evaluate best injection timing and injection pressure with Honge methyl ester and cotton seed oil methyl ester [13]. A toroidal re-entrant combustion chamber of the engine was selected. The injection pressure was varied from 220 to 260 bar and the injection timing of 19°, 23° and 27° BTDC was executed at compression ratio of 17.5. At injection timing of 19° BTDC and injection pressure of 240 bar biodiesel showed best BTE. Honge methyl ester showed better performance compared to cotton seed methyl ester. Further the injector of six holes with 0.2 mm orifice diameter yielded better results compared to 0.25 mm and 0.3 mm orifice diameter. It showed lower brake thermal efficiency for the biodiesel fuel. Smaller orifices have shorter ignition delay which reduces heat loss and time loss resulting in higher brake thermal efficiency. The NO_x emissions were lower whereas the HC and CO emissions were higher with 0.2 mm orifice diameter for the biodiesel fuel.

Tests were conducted for three injection pressures 180, 200, 220 bar for four different loads at 2200 rpm with methanol blended diesel from 0–15% with an increment of 5% [14]. The original pressure of the engine was 200 bar. It was reported that at decreased pressure of 180 bar the heat release rate, combustion efficiency, NO_x emissions, CO_2 emissions and peak cylinder pressure decreased whereas CO, HC and smoke number increased. On the other hand, smoke number, unburned hydrocarbon, and carbon monoxide emissions reduced with 220 bar, and peak cylinder pressure, heat release rate, combustion efficiency, nitrogen oxides and carbon dioxide emissions increased at all loads.

Exhaust emissions and BSFC with biodiesel blends (Name not specified) on diesel engine with different injection pressures at four different loads were studied [15]. All the biodiesel blends showed less CO emissions at lighter loads and high emissions at full loads. It was also noted that the CO emissions and HC emissions decreased with the increase in the biodiesel percentage. At 50 kPa constant load the CO emissions, HC emissions and smoke opacity decreased with the increasing injection pressure. As the engine load increased the NO_x emissions increased for all the blends. There was a rise in the NO_x emissions with the increase in the percentage of the biodiesel. With the increase in injection pressures, blends with high percentage of biodiesel showed less BSFC.

Optimization of injection pressure and compression ratio can improve engine performance [16]. The pressures selected were (150 bar, 200 bar, 250 bar) and compression ratios selected for the test were 16, 17, and 18. On comparing all the combination of pressures and compression ratio it was found that at 250 bar injection pressure and 18 compression ratio gave the highest brake thermal efficiency and lowest brake specific fuel consumption for the Jatropha methyl ester. However the HC and exhaust temperature increased with this combination whereas the smoke and CO emissions reduced. No change was observed with the NO_x emissions at higher pressures. For all the combination of pressure and compression ratio the Jatropha methyl ester showed better results than neat diesel both in case of performance as well as emissions.

In a similar study higher BSFC with linolenic linseed oil methyl ester at higher pressure of 240 bar was reported [17]. Moreover the thermal efficiency improved at 240 bar with some increased emission in NO_x. This may due to the changes in the shape of fuel spray which results in shortening of the length of the spray resulting in higher dispersion and higher spray tip penetration. This ultimately lead to better combustion.The ignition delay was reported to be lower at higher injection pressures as compared to diesel and at full load the peak pressure was the highest. The combustion analysis shows that, the ignition delay is lower at higher injection pressures compared to diesel. Peak pressures increased with the increase in the load for all injection pressures for the biodiesel as well as diesel. Smoke was lower than diesel at all loads for the biodiesel.

5. Conclusion

Fuel reactivity influenced simultaneous reduction in NO_x and soot emissions and increased dwell period, reduced particulate matter for diesel fuel. Multiple injection strategies are effective only with certain adjustment with injection rate as well as the duration of dwell period. Quantity of fuel in the first and second injection also greatly influences brake thermal efficiency and smoke emissions.

Split injections with biodiesel has mitigated NO_x emissions but at the cost of engine performances. Split injections with higher injection pressure reduced soot, HC and CO emissions at the cost of NO_x emissions. Higher injection pressures with biodiesel did not show consistent results with NO_x emissions however when coupled with EGR strategies, it has depicted significant reduction in NO_x emissions.

Author details

Jayashri N. Nair
VNR Vignana Jyothi Institute of Engineering and Technology, Hyderabad, India

*Address all correspondence to: jayashri@vnrvjiet.in

References

[1] Nehmer DA, Reitz RD., Measurement of the Effect of Injection Rate and Split Injections on Diesel Engine Soot and NOx emissions. SAE Transactions, pp.1030-1041, (1994)

[2] Akiba S, Keiya N, Xinyun Z., Effect of Split Injection Pattern on Combustion and Emission Characteristics of DI Diesel Engine. Marine Engineering. Journal of the Japan Institution of Marine Engineering. 2011; 46:441-47. DOI: 10.5988/jime.46.441

[3] Krishna P, Babu AK, Singh AP, Raj AA. Reduction of NOx in a Diesel Engine Using Split Injection Approach. Journal of Engineering Science and Technology.2015; 10:552-70.

[4] Kim, M.Y., Yoon, S.H. and Lee, C.S., Impact of Split Injection Strategy on the Exhaust Emissions and Soot Particulates from a Compression Ignition Engine Fuelled with Neat Biodiesel. Energy & Fuels. 2008; 22:1260-1265. DOI: 10.1021/ef700537w

[5] Iwazaki, K., Amagai, K. and Arai, M., Improvement of Fuel Economy of an Indirect Injection (IDI) Diesel Engine with Two-stage Injection. Energy. 2005; 30: 447-459. DOI: 10.1016/j.energy.2004.05.009

[6] Cowart, Jim, Dianne Luning Prak, and Len Hamilton. The Effects of Fuel Injection Pressure and Fuel Type on the Combustion Characteristics of a Diesel Engine. In ASME 2014 Internal Combustion Engine Division Fall Technical Conference, American Society of Mechanical Engineers. 2014. DOI: 10.1115/1.4027408

[7] Abdullah N, Tsolakis A, Rounce P, Wyszinsky M, Xu H, Mamat R. Effect of injection pressure with split Injection in a V6 diesel engine. SAE Paper NO. 2009-24-0049. DOI: 10.4271/2009-24-0049.

[8] How, H.G., Masjuki, H.H., Kalam, M.A. and Teoh, Y.H., Influence of Injection Timing and Split Injection Strategies on Performance, Emissions, and Combustion Characteristics of Diesel Engine Fuelled with Biodiesel Blended Fuels. Fuel. 2018; 213: 106-114. DOI: 10.1016/j.fuel.2017.10.102

[9] Zhang, Y. and Boehman, A.L., Impact of Biodiesel on NOx Emissions in a Common Rail Direct Injection Diesel Engine. Energy & Fuels. 2007: 2003-2012. DOI: 10.1021/ef0700073

[10] Suryawanshi, J.G. and Deshpande, N.V., Effect of Injection Timing Retard on Emissions and Performance of a Pongamia Oil Methyl Ester Fuelled CI Engine (No. 2005-01-3677). SAE Technical Paper 24 Oct 2005; DOI: 10.4271/2005-01-3677.

[11] Park SH, Youn IM, Lee CS., Influence of Ethanol Blends on the Combustion Performance and Exhaust Emission Characteristics of a Four-cylinder Diesel Engine at various Engine Loads and Injection Timings. Fuel. 2011; 90:748-55. DOI: 10.1016/j.fuel.2010.08.029

[12] Lahane S, Subramanian KA., Effect of Different Percentages of Biodiesel–diesel Blends on Injection, Spray, Combustion, Performance, and Emission Characteristics of a Diesel Engine. Fuel.2015; 139: 537-45. DOI: 10.1016/j.fuel.2014.09.036

[13] Khandal, S.V., Banapurmath, N.R. and Gaitonde, V.N., Different Injection Strategies to Enhance the Performance of Diesel Engine Powered with Biodiesel Fuels. European Journal of Sustainable Development Research. 2017; 1-5. DOI: 10.20897/ejosdr.201715

[14] Canakci, M., Sayin, C., Ozsezen, A.N. and Turkcan, A., Effect of Injection Pressure on the Combustion,

Performance, and Emission Characteristics of a Diesel Engine Fuelled with Methanol-blended Diesel Fuel. Energy & Fuels. 2009; 23:2908-2920. DOI: doi.org/10.1021/ef900060s

[15] Gumus, M., Sayin, C., & Canakci, M. The Impact of Fuel Injection Pressure on the Exhaust Emissions of a Direct Injection Diesel Engine Fuelled with Biodiesel–diesel Fuel Blends. Fuel. 2012; 95: 486-494. DOI: 10.1016/j.fuel.2011.11.020

[16] Jindal, S., Nandwana, B.P., Rathore, N.S. and Vashistha, V., Experimental Investigation of the Effect of Compression Ratio and Injection Pressure in a Direct Injection Diesel Engine Running on Jatropha Methyl Ester. Applied Thermal Engineering. 2010; 30:442-448. DOI: 10.1016/j.applthermaleng.2009.10.004

[17] Puhan, S., Jegan, R., Balasubbramanian, K. and Nagarajan, G., Effect of Injection Pressure on Performance, Emission and Combustion Characteristics of High Linolenic Linseed Oil Methyl Ester in a DI Diesel Engine. Renewable energy. 2009; 34:1227-1233. DOI: 10.1016/j.renene.2008.10.001

Chapter 7

Combustion and Emissions of Gasoline Compression Ignition Engine Fuelled with Gasoline-Biodiesel Blends

Yanuandri Putrasari and Ocktaeck Lim

Abstract

A gasoline compression ignition (GCI) engine was proposed to be the next generation internal combustion engine for gasoline. The effect of exhaust gas recirculation (EGR) and intake boosting on combustion and emissions of GCI engine fueled with gasoline-biodiesel blends by partially premixed compression ignition (PPCI) combustions are investigated in this study. Tests were conducted on a single-cylinder direct-injection CI engine, with 5% by volume proportion of biodiesel in gasoline fuel blends. Engine control parameters (EGR rate, intake boosting rate, and various injection strategies) were adjusted to investigate their influences on combustion and emissions of this GCI engine. It is found that changes in EGR rate, intake boosting pressure and injection strategies affect on ignition delay, maximum pressure rise rate and thermal efficiency which is closely tied to HC, CO, NO_x and smoke emissions, respectively.

Keywords: gasoline, fuel, GCI, thermal efficiency, emission

1. Introduction

The stricter limitations of vehicle emission regulation especially for compression ignition (CI) engines with petroleum diesel fuel motivates many researchers to explore the utilization of low volatile and alternative fuels for CI engines to obtain high efficiency, but produce lower emissions or so-called low temperature combustion (LTC). Due to the low volatility and short ignition delay of diesel fuel, CI engines produce high nitrogen oxides (NO_x) and soot emissions. To obtain LTC combustion, which is increase engine efficiency, and improve exhaust emissions, a variety of combustion methods in CI engines have been investigated such as homogeneous charge compression ignition (HCCI), PPCI, multiple premixed compression ignition (MPCI), etc. HCCI engine is one method that potentially to achieve an advanced LTC, which possible to produce low particulate matter (PM) and NOx emissions to replace conventional diesel engine combustion [1]. However, many technical problems of HCCI engine strategy must be solved before released to the market. The maximum load limitation due to the surplus of pressure rise rate (PRR) and engine knocking phenomenon [2, 3] have been major obstacle as long as HCCI combustion influenced by fuel type and air fuel mixture quality.

Recently, several studies have shown that gasoline and some other fuels with low cetane number and higher volatility are potentially advantages for low temperature combustion in CI engine, which is popular as gasoline compression ignition engine (GCI) [4–19]. At the beginning, GCI was proposed to exploit the benefit of high volatility and long ignition delay of gasoline fuel and high ratio of CI engine to obtain both high engine efficiency and low exhaust emission [20]. Later on, the concept of GCI was improved with PPCI concept by injecting fuel in the compression stroke, then mixture stratification is formed before combustion event [21]. The PPCI concept may extend load range at the same time maintaining high thermal efficiency, low NOx and soot emissions. Even though, the pressure rise rate (PRR) is remaining too high at the high load operation. The most advanced injection strategy on GCI is MPCI which has purpose to control combustion noise by manage the injection and combustion process in an order as spray – combustion – spray – combustion [22, 23]. By using MPCI mode, the acceptable pressure rise rate can be achieved and extended load range can be obtained.

There are many challenges in the implementation of GCI mode in CI engines. Several techniques are used to realize the utilization of gasoline fuel in CI engines including PPCI, MPCI and other various parameters. The start of injection (SOI) timing has effects on the balance of GCI combustion based on the study by Kodavasal et al., [24]. The study indicated that, the perfect combustion of a GCI engine was earned with start of injection (SOI) timing at -30° after top dead center (ATDC) and misfired was happened at SOI advanced than -42° ATDC. More advance SOI has potentially to solve the exhaust gas deterioration by minimizing NO_x and keeps the high performance of LTC. However, more knocking, ringing, HC and CO will occur due to improper combustion. The multiple-injection strategy has been known as the solution for the NOx/soot and decrease of combustion noise with maintaining low fuel consumption in CI engine. The application of this method with higher volatility fuel which are suitable for well-premixed or properly stratified mixture prior to the ignition to reduce both NO_x and soot emissions while maintaining high efficiency compared to conventional diesel CI engines was conducted previously [21, 25]. Meanwhile, the double injection strategy for pure gasoline fuel in GCI engines successfully reduced MPRR and NOx levels in half of the single-injection [17]. However, the IMEP and fuel economy were decreased. Furthermore, the significant increasing of the CO and soot emissions were also happened, even though the levels less than conventional diesel CI engine.

Blending gasoline fuel with certain percent of biodiesel in CI engines is the one way to obtain the good combustion and emissions results. Biodiesel is proven to be appropriated as a substitution fuel for CI engines [11, 26]. Furthermore, biodiesel has evidently decreasing engine exhaust pollutant [27, 28], due to the high content of oxygen which important to minimize the soot development in combustion process [29]. The implementation of gasoline-biodiesel blends in GCI engine with single and double injection modes has been studied previously [30–32]. The effect of biodiesel-gasoline blends on GCI combustion using 5% and 10% biodiesel was studied by Adams [30], which focused on reducing required intake temperature and utilization of split injection. To overcome the auto ignition difficulty of gasoline fuel without modification on intake temperature in GCI engine, the authors using a high compression ratio around 19.5 and various SOI of single injection mode in the previous study [31].

Previous studies have presented detailed analysis and discussion of the combustion and emission characteristics of GCI for PPCI or MPCI modes, by fueled with gasoline-diesel blends or gasoline biodiesel blends using direct injection GCI concept [4, 7–10, 12, 30–34]. However, the combustion and emission characteristics of CI engines are also influenced by various other factors, such as fuel injection strategy, initial conditions and its combustion modes. Furthermore, based on previous study [30–32], the maintaining of high efficiency and emission reduction of GCI engine fueled with gasoline-biodiesel blends still challenging and need more to be optimized especially for its NOx and soot emissions. Since the auto-ignition sensitivity of gasoline fuel is influenced by several factors such as in-cylinder equivalence ratio, intake dilution, intake temperature and pressure, it is potentially to utilize EGR and boosting in GCI combustion. To overcome the load operation limit and adjusting heat release subject to engine speed by delaying the combustion phasing, EGR was used [35]. The high-load operation can be achieved without knocking by using EGR, in which increases of specific heats capacity and minimize of oxygen (O_2) concentration in the chamber promotes longer ignition delay and shifted combustion phasing far away from after top dead centre [36–39]. But, surplus of EGR supply leads to the decreasing of power and resulted more CO and HC emissions. Thus, boosting was utilized to increase the operating load simultaneously encourage fuel ignition reactivity. Furthermore, adjusting the CA50 is also necessary when boosting is operated. Because, the more intake charges mass leads to the high intensity of knocking due to the higher of pressure rise [40–42]. Suitable method of EGR and boosting was proven potentially to extend CI engine 12]. Thus, it is also potential to be used in GCI engines. The basic and control mechanism of EGR with boosting on GCI gasoline-biodiesel auto-ignition should be able to explain the relatively wide ranges of operating parameters. Thus, complementary experimental works are conducted to achieve a better understanding on the combustion process and emission characteristics of GCI engine fueled with gasoline-biodiesel blends. Information about the effects of EGR and boosting on GCI engines using gasoline-biodiesel blends are essential for advancing the theory and contributions to successfully implement gasoline in CI engines and biofuel into the transportation sector.

The objective of this study was to determine the effects of EGR and boosting on the combustion and emissions of a GCI engine fueled with gasoline-biodiesel blends. To obtain a clear and comprehensive analysis of the effect of various EGR and boosting rates on combustion and emissions of GCI engine the same basic energy input of injected fuels was used for comparing the various parameters. The PPCI combustion modes for the gasoline-biodiesel blend were utilized. Modification of several initial conditions, such as intake, oil, and coolant temperatures are also conducted. The combustion characteristics of cylinder pressure, heat release rate, ignition delay, and emission characteristics are analyzed accordingly as the focus of this study.

2. Method

2.1 Test system

The experimental study was conducted using a single-cylinder, four-stroke, direct injection, water-cooled, naturally aspirated diesel engine with 498 cm^3 of displacement and four-valve SOHC. **Figure 1** shows the schematic diagram of the test engine and measurement setup. The engine specifications are listed in **Table 1**. A standalone supercharger made by Engine Tech, a Korean local company, was used to supply the intake boosting. A conventional EGR system was used, in which the line is routed directly from exhaust manifold to the intake manifold. The engine was connected to the test system which is a 57 kW Dynamometer (Elin AVL Puma MCA325MO2). A data acquisition system (Dewetron DEWE-800-CA) in combination with an encoder (Autonics E40S8–1800-3-T-24), a pressure transducer (Kistler

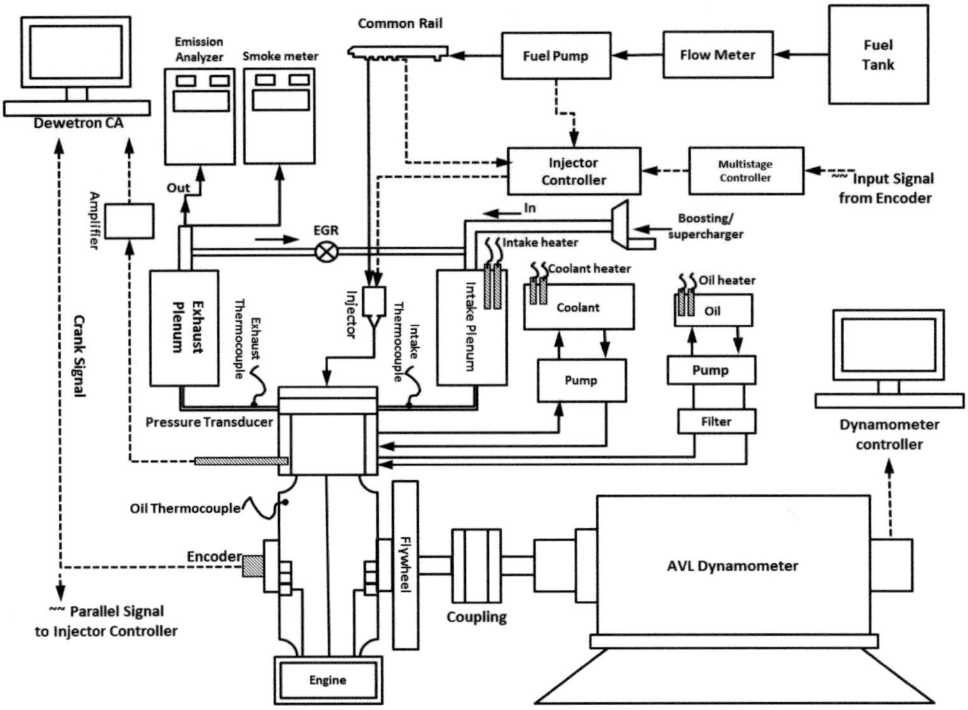

Figure 1.
Schematic diagram of the engine and measurement system setup.

Engine Parameters	Value
Displacement	498 cm^3
Bore	83 mm
Stroke	92 mm
Compression Ratio	19.5
Con. Rod Length	145.8 mm
Crank Radius	43.74 mm
Valve System	4-valve SOHC
Fuel System	Electronic Common Rail

Table 1.
Engine specifications.

6056A), and an amplifier (Kistler 5018) were used to obtain the combustion data. The fuel was injected to the combustion chamber using a Bosch seven-hole injector. A multi-stage injection engine controller (*Zenobalti:* ZB-8035, ZB-5100, ZB-100) was used to adjust the combustion strategy. Meanwhile, the temperatures of air intake, engine coolant, and lubricant oil were controlled by using separate temperature controller. The exhaust emissions, including unburned total hydrocarbon, carbon monoxide, and NOx were measured using a Horiba MEXA-7100DEGR. AVL 415 smoke meter was used for soot emission measurement. Some thermocouple K and RTD types were installed on the certain part of the engine to measure including the intake, oil, coolant and exhaust temperature.

2.2 Test fuels preparation

Two fuels, which are diesel and gasoline-biodiesel blend were used in this study. The baseline fuels utilized in this study were commercial gasoline (GB00), neat diesel (D100) and pure soya bean biodiesel (B100). The chemical composition information of soya bean vegetable oil is given in **Table 2**. Biodiesel (5% by volume) and gasoline were blended and labeled as GB05. To maintain the homogeneity, the mixing process was conducted about 10 minutes then immediately used for experiment. The physical properties of baseline fuels and GB05 are given in **Table 3**.

2.3 Operating conditions

The engine was operated at stable condition with fixed 1200 rpm. An injection pressure of 70 MPa was used for PPCI. Single injection timing at 40 °CA BTDC was adopted and set for PPCI combustion mode. The total energy input of injected fuel was set at around 26 mg/cycle. The initial parameters of intake temperature, oil temperature, and coolant temperature were maintained at 85°C, 75°C, and 65°C, respectively. The reason why the intake temperature was maintained at 85°C is to promote the autoignition of the fuel easier. As already known that GB05 more likely as a pure gasoline which low autoignition characteristics. The homogeneous hot

Fatty Acid	System Name	Structure	Formula[a]	Composition (wt %)
Myristic	Tetradecanoic	14:0	$C_{14}H_28O_2$	0
Palmitic	Hexadecanoic	16:0	$C_{16}H_{32}O_2$	12
Stearic	Octadecanoic	18:0	$C_{18}H_{36}O_2$	3
Arachidic	Eicosanoic	20:0	$C_{20}H_{40}O_2$	0
Behenic	Docosanoic	22:0	$C_{22}H_{44}O_2$	0
Lignoceric	Tetracosanoic	24:0	$C_{24}H_{48}O_2$	0
Oleic	cis-9-Octadecenoic	18:1	$C_{18}H_{34}O_2$	23
Linoleic	cis-9,cis-12-Octadecadienoic	18:2	$C_{18}H_{32}O_2$	55
Linolenic	cis-9,cis-12, cis-15-Octadecatrienoic	18:3	$C_{18}H_{30}O_2$	6
Erucic	cis-13-Docosenoic	22:1	$C_{22}H_{42}O_2$	0

xx:y indicates xx carbons in the fatty acid chain with y double bonds.

Table 2.
Chemical composition of soya bean vegetable oil.

Test Item	Unit	Test Method	Gasoline	GB05	B100	D100
Heating Value	MJ/kg	ASTM D240:2009	45.86	45.32	39.79	45.93
Kinematic Viscosity (40°C)	mm^2/s	ISO 3104:2008	0.735	—	4.229	2.798
Lubricity	mm	ISO 12156-1:2012	548	290	189	238
Cloud Point	°C	ISO 3015:2008	−57	−37	3	−5
Pour Point	°C	ASTM D6749:2002	−57	−57	1	−9
Density (15 °C)	kg/m^3	ISO 12185:2003	712.7	722.3	882.3	826.3

Table 3.
Physical properties of the fuels.

EGR and air mixture were applied in this study with 0%, 20% and 50% of flow rates by using a pair of gate valve. The EGR ratio was calculated using Eq. 1 as follows.

$$EGR\% = \frac{m_E}{m_E + m_i} \times 100\% \tag{1}$$

where m_E and m_i are the mass of EGR and intake fresh air, respectively.

The air boosting were set at 0.1 and 0.12 MPa in the intake manifold. The more detail engine operating parameters and injection strategies are presented in **Tables 4** and 5, respectively. The data of 100 consecutive cycles such as in-cylinder pressure was recorded for combustion analysis.

The analysis and discussion was performed on several engine parameters such as in cylinder heat release rate, temperature, peak of pressure rise rate, IMEP, COV of IMEP, knocking/ringing intensity, thermal efficiency and combustion efficiency. Rate of heat release was calculated using Eq. 2.

$$\frac{dQ}{d\theta} = \frac{1}{\gamma - 1} V \frac{dp}{d\theta} + \frac{\gamma}{\gamma - 1} p \frac{dV}{d\theta} \tag{2}$$

Where, γ is the specific heat ratio, V is the instantaneous cylinder volume, and p is the cylinder pressure. The normal and suitable value of γ for a CI engine is 1.3. The in-cylinder pressure and volume data were used to calculate the in-cylinder temperature using ideal gas law, as shown in Eq. 3.

$$T = \frac{p.V}{n.R} \tag{3}$$

where p is for pressure, V for volume, n is the amount of substance, and R is the gas constant.

Parameter	Diesel/GCI
Speed (rpm)	1200
Inj. Pressure (MPa)	70
Injection strategy	PPCI
Inj. Quantity (mg)	26
T intake (°C)	85
T oil (°C)	75
T coolant (°C)	65
EGR (%)	0, 20 and 50
Intake boosting (MPa)	0.1 and 0.12

Table 4.
Operating parameters.

Combustion modes	Injection strategies	Injection timing and duration	Injected fuel	
			D100	GB05
PPCI	Single	40 °CA BTDC (1000 μs)	(26 mg)	

Table 5.
Injection strategies.

Furthermore, the emission of CO, HC, NO_x and particulate matter (smoke) were also discussed and analyzed in detail.

3. Experimental results and discussion

The main purpose of this experiment is to improve the efficiency and emission characteristics of CI engine fueled with gasoline-biodiesel blends using GCI mode. To replace the utilization of diesel fuel with gasoline fuel in CI engine due to high demand of diesel fuel in the market, thus a small amount of biodiesel (5%) was added as the lubricity improver to overcome the wear problems in the fuel system. The performance results were compared with only pure diesel fuel, because the basic of the engine is diesel engine. The engine was run on single injections mode (PPCI) at 1200 rpm with various EGR rate (0%, 20% and 50%) and intake boosting 0.1 to 0.12 MPa to investigate the effect of EGR and boosting on performance, combustion and emissions. The performance, combustion and emissions characteristics data was analyzed and presented graphically for in-cylinder pressure, temperature, HRR, ignition delay, MPRR, PPRR, IMEP, thermal efficiency and its emissions including HC, CO, NO_x and smoke opacity.

3.1 Effect of EGR and PPCI injection strategy

The total fuel consumption per cycle in PPCI mode is maintained at 26 mg per cycle and single injection timing at 40 °CA BTDC. The others engine operating conditions i.e. air intake, engine coolant and engine oil temperatures were set at 358 K, 338 K and 348 K, respectively. Meanwhile, the fixed intake pressure 0.1 MPa and various EGR rates for 0%, 20% and 50% were used to characterized the effect of EGR and PPCI injection strategy on combustion and emissions of GCI engine fueled with gasoline-biodiesel blends.

Figure 2 shows the in-cylinder pressure, temperature and HRR of PPCI mode at various EGR rate for 0%, 20% and 50%, and fixed intake boosting 0.1 MPa. It can be seen from the figure that CI engine fueled with diesel fuel reveal the decreasing in-cylinder pressure when the EGR rate is increase. Similar with the diesel fuel, gasoline-biodiesel blend also indicates the same trend when EGR rate increase the in-cylinder pressure decrease. The in-cylinder temperature for diesel fuel decreasing as the trend of in-cylinder pressure when EGR rate increase. However, the in-cylinder temperature trends of gasoline-biodiesel blend show that EGR 50% lead to the highest value among the others EGR rates. Observing at heat release rates curves, it is seen that the heat release process of both diesel and gasoline-biodiesel blends fuels show a marked two-stage ignition. The first stage ignition of diesel fuel consistently higher than 20 J/deg., even though all of the curves reveal decreasing trends for various increasing EGR rate. Meanwhile, the first stage ignitions from gasoline-biodiesel blends are very low for all various EGR rates, and it is almost very difficult to be recognized. The highest peak of heat release rate can be obtained from gasoline-biodiesel blends with 50% EGR rate. The highest peak of heat release rate can be used to determine that the excessive pressure rise rate is happened. The excessive of PRR means that the combustion is not stable or some time when in the high load condition, the rapid pressure rise rate can result in heavy knocking operation.

Figure 3 shows the effect of EGR on ignition delay when engine operated using PPCI mode. The higher EGR rate results the longer ignition delay for both of diesel and gasoline-biodiesel blends. However, it can be observed that gasoline-biodiesel blends lead to the much longer ignition delay compared to diesel fuel in every EGR

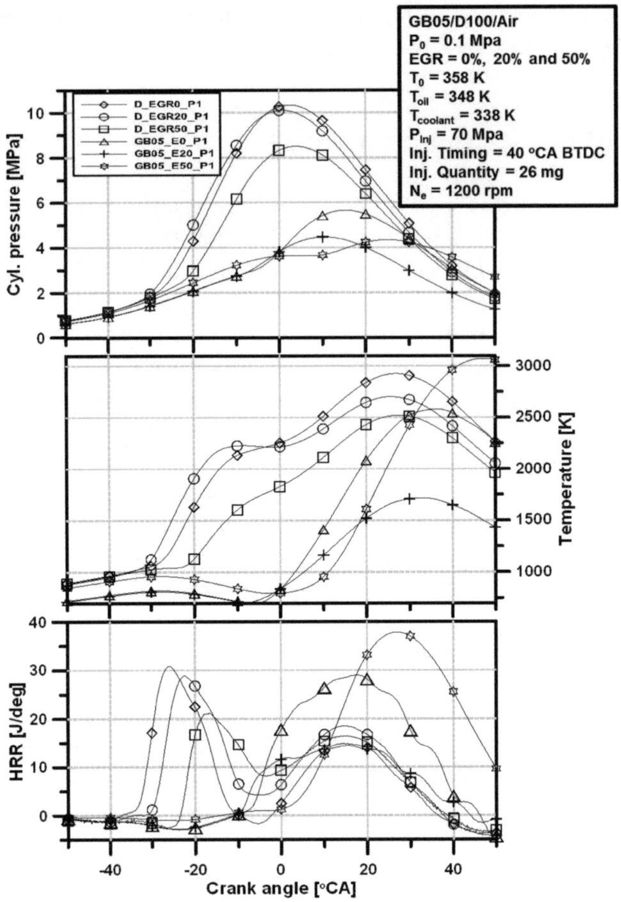

Figure 2.
Effect of EGR on cylinder pressure, temperature and HRR of PPCI mode.

rate variations. This condition is the advantage of gasoline fuel, which is longer ignition delay due to high volatile and low cetane number, thus there is a possibility of complete mixing period before combustion occurred. However, the longer ignition delay caused the shifted of maximum in-cylinder pressure far away form TDC which can reduce the performance of the engine. To overcome this condition the earlier injection timing can be applied among many other solutions.

Figure 4(a) shows the effect of EGR on maximum of in-cylinder pressure and (b) peak of pressure rise rate of PPCI strategy. The higher of EGR rate generate the lower in-cylinder pressure maximum and lower the maximum pressure rise rate for both diesel fuel and gasoline-biodiesel blends. Similarly, the increasing of EGR rates also reducing the maximum of pressure rise rate for both diesel fuel and gasoline-biodiesel blends. This condition happened due to the slowdown of combustion process. One of the reasons when utilizing EGR to slowing down of combustion process is the concentration of O_2 is lowered and the concentrations of CO_2 and H_2O unintentionally increased. Therefore, this slows down the reactions in the oxidizing direction and speeds up the reactions of reduction process direction.

The effect of EGR on IMEP of GCI engine using PPCI strategy is presented in **Figure 5**. The increasing of EGR rates does not give any effect on IMEP of GCI engine fueled with diesel fuel. However, the 50% EGR rate results the highest IMEP value for gasoline-biodiesel blends, even much higher if compared with diesel fuel that is almost 1.0 MPa. Related to the IMEP value, the engine efficiencies especially

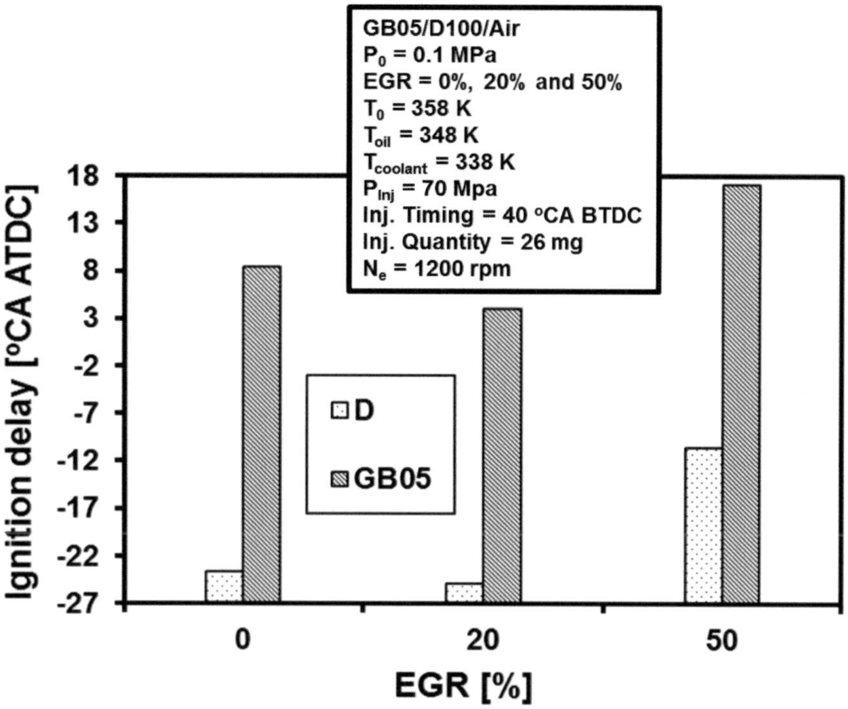

Figure 3.
Effect of EGR on ignition delay of PPCI mode.

indicated thermal efficiency also can be calculated by using it derivative that is indicated power/work. The effect of various EGR rates on indicated thermal efficiency of GCI engine using PPCI strategy can be seen in **Figure 6**. It can be seen that by increasing EGR rate the value of indicated thermal efficiencies are decreased for both of diesel and gasoline-biodiesel blends. The 50% EGR rate for diesel fuel lead to a little increasing value of indicated thermal efficiency is compared with 20% of EGR rate. However, it caused the significant drop value of indicated thermal efficiency in case of gasoline-biodiesel blends fuel.

Effect of EGR rates on CO emission of GCI engine using PPCI mode can be observed on **Figure 7**. All variation of EGR rates showed that CO emission of gasoline-biodiesel blends are lower than diesel fuel due to the volatile properties of gasoline and higher oxygen content of biodiesel, which make more complete mixing and produce more perfect combustion. However, in general, the increasing of EGR rates caused no different of CO emission for both diesel and gasoline-biodiesel blends fuels. A little decreasing value of CO emission was only happened on GCI engine fueled with gasoline-biodiesel blends when running on 50% EGR rate.

Figure 8 shows the effect of various EGR rate on HC emission of GCI engine running on PPCI strategy. As like the trend of CO emission, HC emission of GCI engine fueled with gasoline-biodiesel blends was also showed a lower value compared to diesel fuel. This condition can be explained also due to the properties of gasoline fuel and the oxygen content of biodiesel. The 20% of EGR rate value gives the lowest effect of HC emission both for diesel and for gasoline-biodiesel blends. Therefore, it is assumed in the PPCI mode the 20% of EGR rate as an optimum value to obtain lowest HC emission.

The NO_x emission and its effect by using various EGR rate on GCI engine using PPCI mode can be seen in **Figure 9**. Normally, the increasing of EGR rates will lead to the lower NO_x emission. However, in this case, for diesel fuel, the 20% of EGR

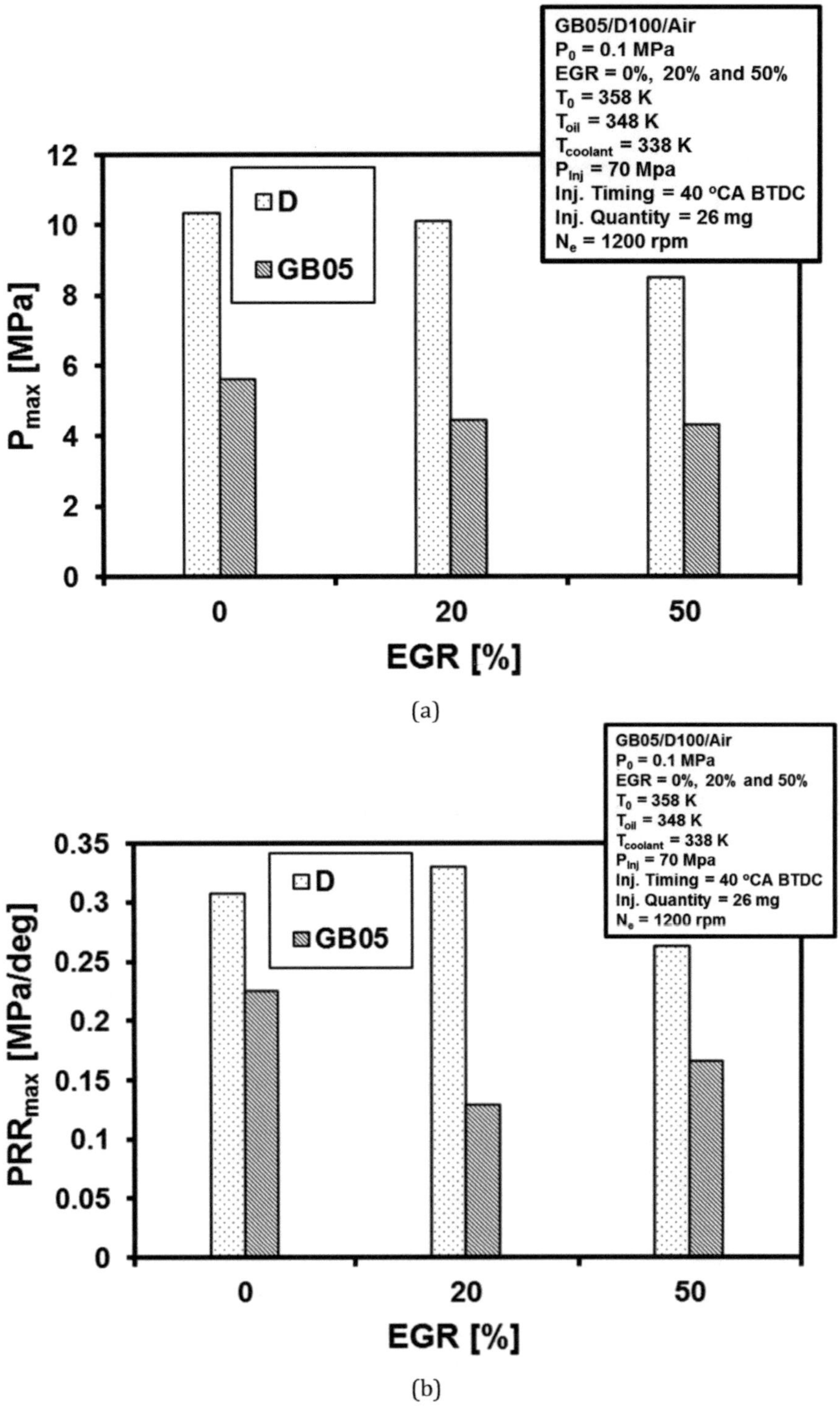

Figure 4.
Effect of EGR on (a) max pressure and (b) peak pressure rise rate of PPCI mode.

rate gives highest NO_x emission. Even though, when 50% EGR was applied the NO_x emission will also decreasing. However, there are no effects of EGR rate variations on NO_x emission of GCI engine fueled with gasoline-biodiesel blends. This

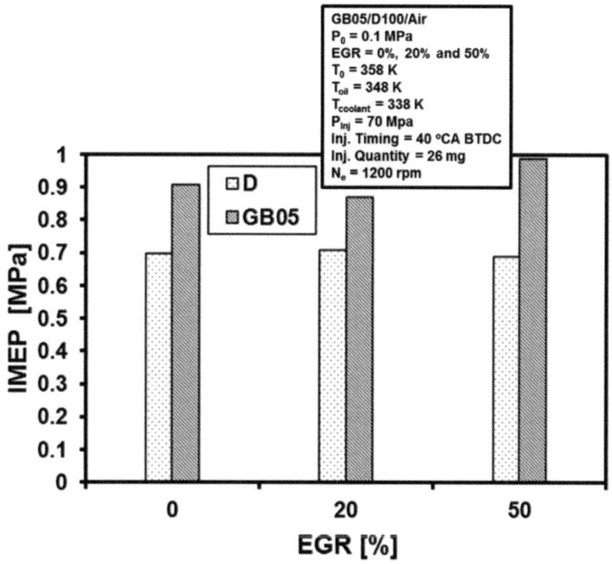

Figure 5.
Effect of EGR on IMEP of PPCI mode.

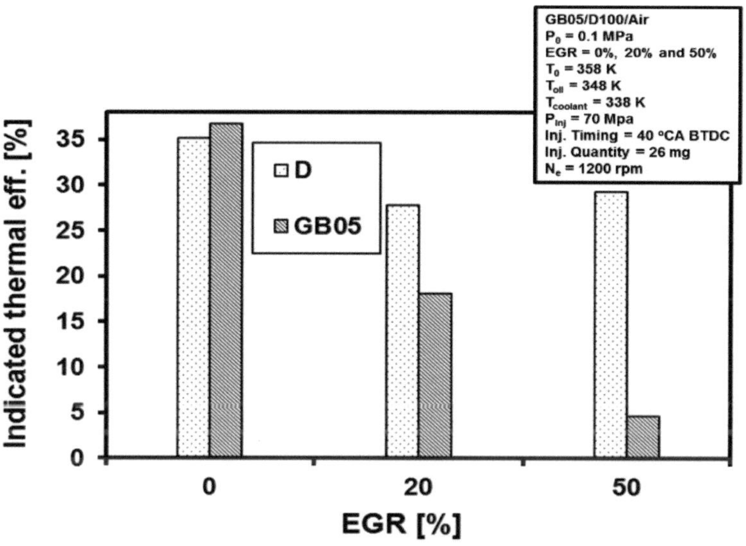

Figure 6.
Effect of EGR on indicated thermal efficiency of PPCI mode.

condition can be seen in the trend of graph that from the three EGR rate variation resulted almost same NO_x emission value.

The smoke emission of CI engine usually contrasts with NO_x emission. When the NO_x higher, the smoke will be a lower and vice versa. The effect of EGR rate variation on the smoke emission of GCI engine can be seen in **Figure 10**. Smoke emission of GCI engine fueled with diesel in the high level for all variation of EGR rate, even when the rate increased. However, the smoke emission of GCI engine using gasoline-biodiesel blends obtain its lowest value when EGR rates at 20%. It can be said that the optimums of EGR rate that can maintain lowest smoke emission while lowest NO_x emission is 20%.

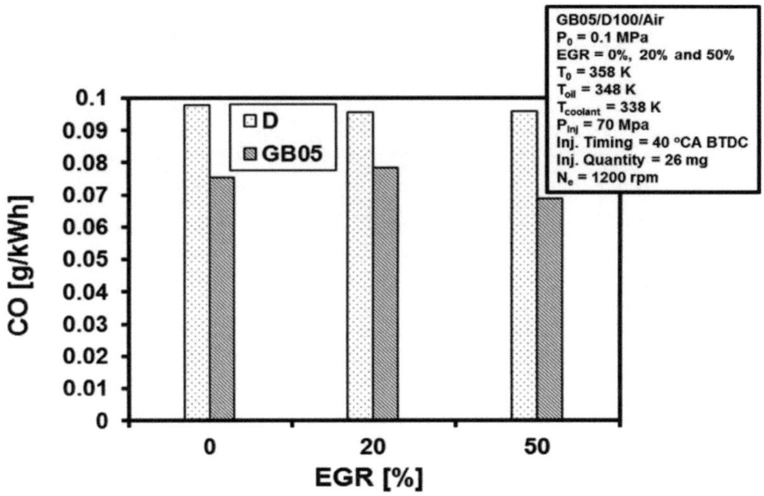

Figure 7.
Effect of EGR on CO emission of PPCI mode.

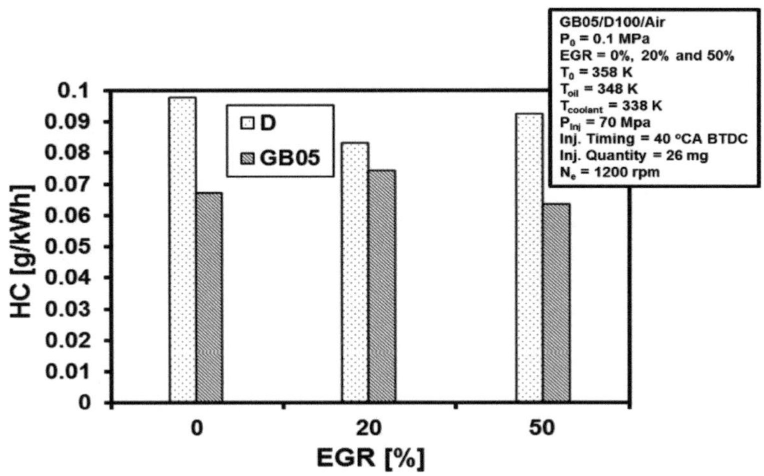

Figure 8.
Effect of EGR on HC emission of PPCI mode.

3.2 Effect of intake boosting and PPCI injection strategy

To understand the effects of intake boosting on GCI engine fueled with gasoline-biodiesel blends on PPCI mode in a simple and easy way, only the 20% of EGR rate was chosen an explained in this study. The intake boosting was set at 0.1 MPa and 0.12 MPa. **Figure 11** shows the effect of boosting on in-cylinder pressure, temperature, and heat release rate of GCI engine fueled with gasoline-biodiesel blends when running on PPCI strategy. Normally found that the increasing of intake boosting rate, increasing the in-cylinder pressure for both diesel fuel and gasoline-biodiesel blends fuel. An ambient pressure of intake boosting gives a higher in-cylinder pressure of GCI engine fueled with diesel compared to gasoline-biodiesel blends. Even, the in-cylinder of gasoline biodiesel-blends with intake boosting 0.12 MPa is lower than diesel fuel with ambient intake boosting. It was also same, that the implementation of 0.12 MPa intake-boosting leads to a higher in-cylinder pressure

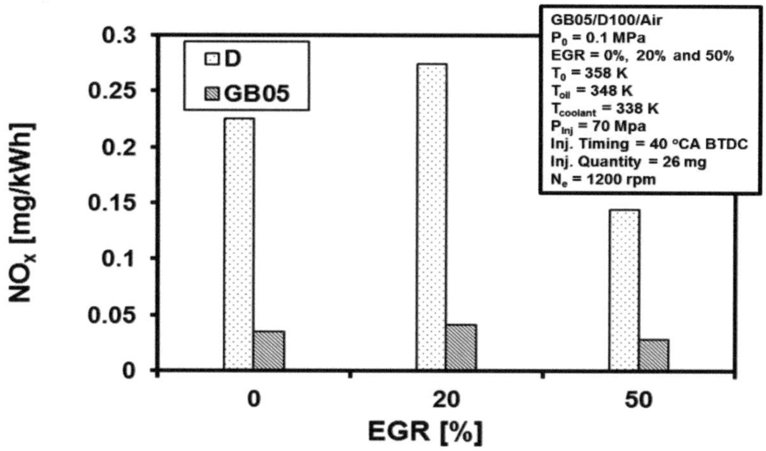

Figure 9.
Effect of EGR on NO_x emission of PPCI mode.

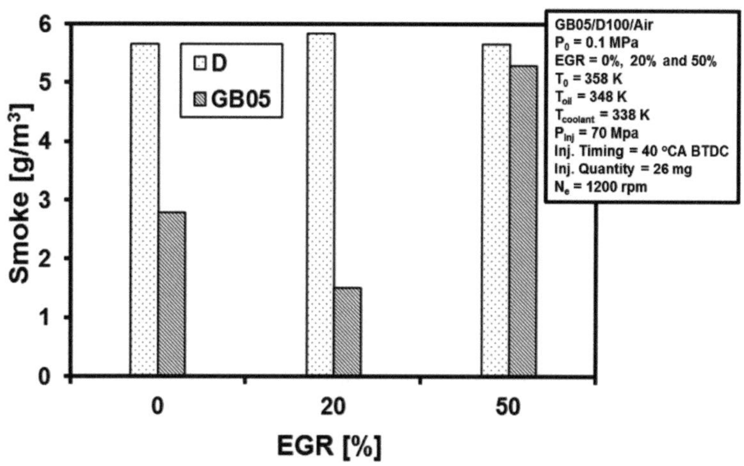

Figure 10.
Effect of EGR on smoke emission of PPCI mode.

of GCI engine fueled with gasoline-biodiesel blends than gasoline-biodiesel with 0.1 MPa intake boosting. To obtain in cylinder pressure of gasoline biodiesel blend at least equal to pure diesel fuel, the higher intake boosting can be applied as long as the engine material supports for high pressure condition and the real engine booster in this case turbocharger can achieves maximum desired pressure. Similar with in-cylinder pressure, the in-cylinder temperature curves show that the highest value is for GCI engine fueled with diesel fuel when intake boosting 0.12 MPa was applied. The lowest in-cylinder temperature, which is below 2000 K, was happened for GCI engine fueled with gasoline-biodiesel blends fuel when using ambient pressure 0.12 MPa. The HRR curves show that the highest value is for GCI engine fueled with gasoline-biodiesel fuel using 0.1 MPa intake boosting. The higher HRR value, the higher-pressure rise rate that can be determines the more unstable engine combustion. The lowest HRR value was obtained from GCI engine fueled with diesel fuel in the ambient pressure condition, which is the most stable combustion.

The effect of intake boosting on ignition delay of GCI engine using PPCI strategy is presented in **Figure 12**. The intake boosting gives effect on the lower ignition

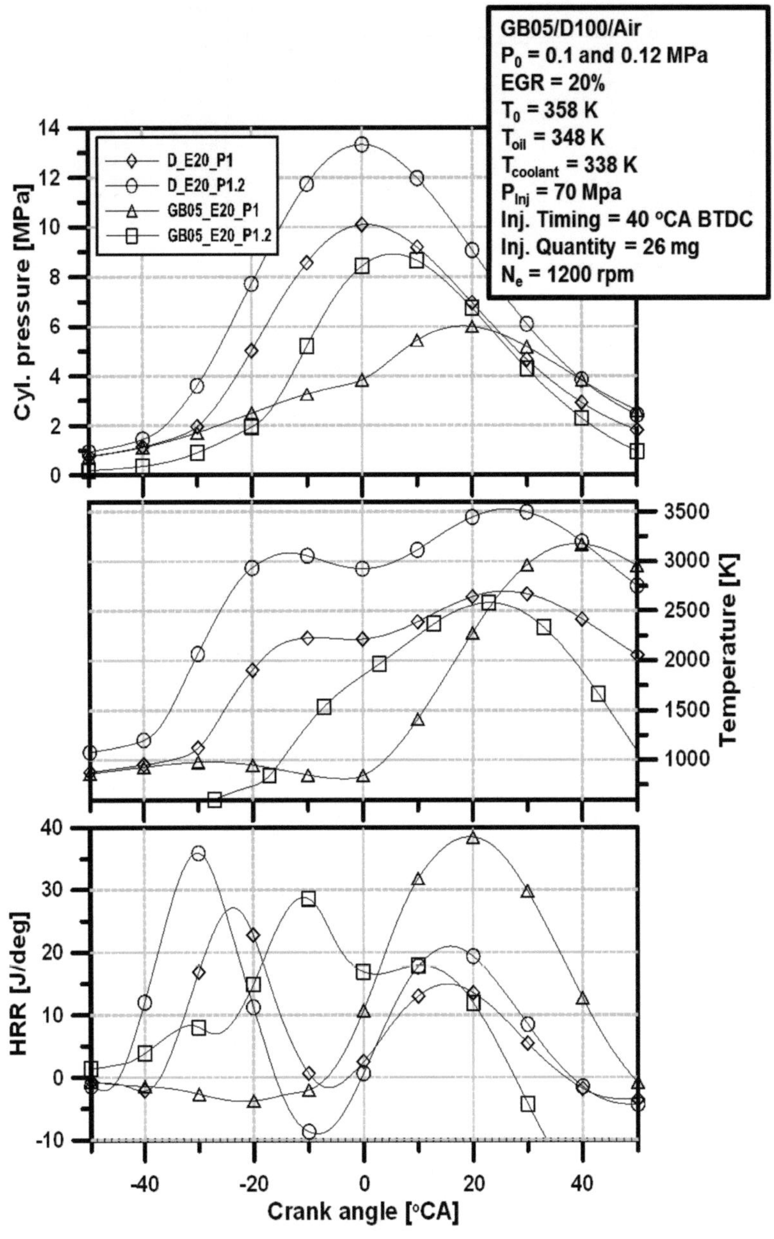

Figure 11.
Effect of boosting on cylinder pressure, temperature, and heat release rate of PPCI mode.

delay for both diesel and gasoline-biodiesel blend fuel. The ambient pressure of intake boosting resulted ignition delay timing for diesel fuel at around 25 °CA BTDC, then the 0.12 MPa intake boosting lead to the slightly earlier of ignition delay timing at around 27 °CA BTDC. Similar trend happened on gasoline-biodiesel fuel, that ambient pressure of intake boosting resulted ignition delay timing at around 11 °CA BTDC, then when 0.12 MPa intake boosting was applied the ignition delay timing also more advanced at around 2 °CA BTDC. The higher volatile and lower cetane number properties of gasoline fuel caused the longer ignition delay timing if compared with diesel fuel. However, the application of intake boosting resulted a shifting of ignition delay timing earlier. The longer ignition delay timing is possible to produce more complete mixing period of air and fuel prior to combustion,

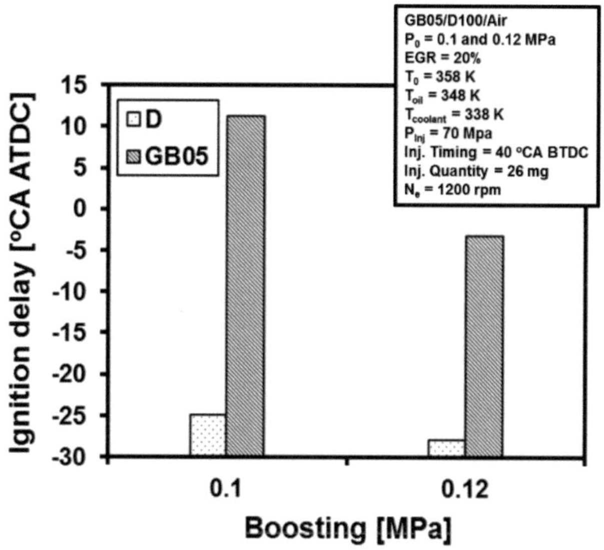

Figure 12.
Effect of boosting on ignition delay of PPCI mode.

however, too long ignition delay timing sometimes caused problem in the engine emission and efficiency.

Figure 13 shows the effect of various intakes boosting on maximum of in-cylinder pressure and its maximum pressure rise rate. A normal condition happened that the increasing intake boosting, the increasing maximum in-cylinder pressure for both diesel and gasoline-biodiesel blends fuels. However, the increasing level of maximum in-cylinder pressure of diesel fuel is much higher than gasoline-biodiesel fuel. It is suspected that intake boosting caused the mixing of air fuel in diesel fuel more optimum than gasoline-biodiesel blend. The increasing of intake boosting leads to the increasing maximum pressure rise rate of GCI engine fueled with gasoline-biodiesel blend in almost same value with diesel fuel. It is mean that the GCI engine running with intake boosting for gasoline-biodiesel blend has an almost similar stability compared with diesel fuel. However, very high-pressure rise rate indicated that the engine in unstable condition.

Effect of various intakes boosting on IMEP of GCI engine fueled with gasoline-biodiesel blends in PPCI strategy can be seen in **Figure 14**. The IMEP of GCI engine fueled with gasoline-biodiesel blend in ambient pressure of intake boosting is higher than when intake boosting is 0.12 MPa. The opposite condition was happened for diesel fuel, which is the IMEP value of GCI engine is higher when 0.12 MPa intake boosting was applied compared with ambient pressure. The condition for IMEP of diesel fuel as the effect of increasing the intake boosting is the normal phenomenon; however, for gasoline-biodiesel blend it is quiet special. This condition suspected by the effect of high volatile and low cetane number of gasolines, which resulted higher-pressure rise rate as shown in **Figure 13**. Fluctuate of in-cylinder pressure may lead to the unstable combustion and resulted the lower IMEP value.

The indicated thermal efficiency of GCI engine using PPCI strategy affected by various intake boosting is presented in **Figure 15**. The indicated thermal efficiency of GCI engine fueled with diesel fuel increased due to the increasing of intake boosting. Similarly, for GCI engine fueled with gasoline-biodiesel blend, even though the IMEP reduced when the intake boosting increased to be 0.12 MPa. This condition, in any case, is expected in the GCI engine fueled with gasoline-biodiesel blend. Furthermore, both for ambient and 0.12 MPa intake boosting showed that

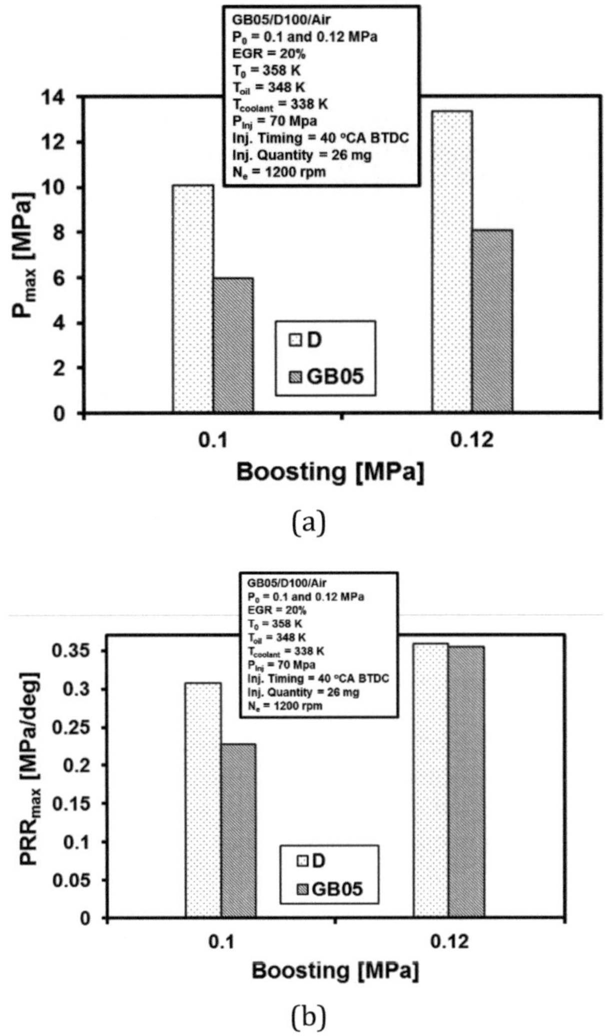

Figure 13.
Effect of boosting on (a) max pressure and (b) peak pressure rise rate of PPCI mode.

the indicated thermal efficiency of GCI engine with diesel fuel is higher than gasoline-biodiesel blend.

Figure 16 shows the effect of intake boosting on CO emission of GCI engine using PPCI strategy. It is already known that the utilization of gasoline-biodiesel blend in GCI engine resulted lower CO emission compared to diesel fuel. Similarly, in the single injection method of PPCI strategy also obtained the lower CO emission of GCI engine fueled with gasoline-biodiesel blend compared to diesel fuel. The increasing of intake boosting from 0.1 to 0.12 MPa in GCI engines gives effect on the decreasing of CO emission for both gasoline-biodiesel blend and diesel fuels. It is suspected due to the combination of 20% EGR and 0.12 MPa of intake boosting, which may lead to the complete combustion.

The effect of intake boosting on HC emission of GCI engine can be observed in **Figure 17**. Similar with the trend on CO emission, the HC emission of GCI engine fueled with of GCI engine with gasoline-biodiesel blend originally is lower than diesel fuel as it can be seen in the ambient intake boosting condition. When the intake boosting increased to be 0.12 MPa HC emission of GCI engine decreased around a half value than 0.1 MPa of intake boosting. For GCI engine fueled with

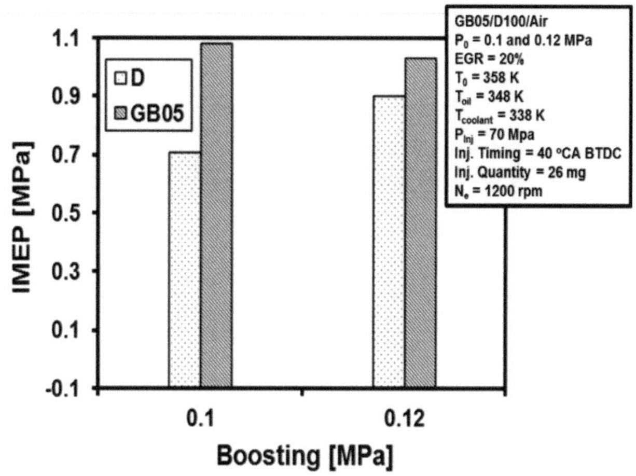

Figure 14.
Effect of boosting on IMEP of PPCI mode.

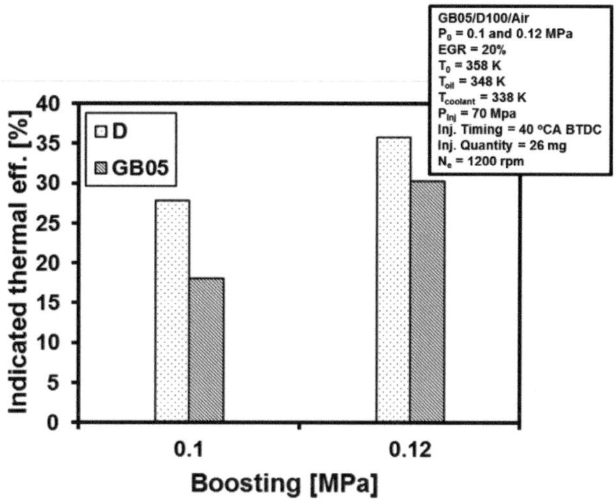

Figure 15.
Effect of boosting on indicated thermal efficiency of PPCI mode.

gasoline-biodiesel blend, it is obtained greatly decreasing of HC emission when the 0.12 MPa of intake boosting applied compared with 0.1 MPa. The decreasing value of HC emission in 0.12 MPa of intake boosting is almost 90% lower from the ambient pressure of intake boosting.

Figure 18 shows the effect of intake boosting on NO_x emission of GCI engine with PPCI strategy. Overall, the NO_x emission of GCI engine fueled with diesel is higher than GCI engine fueled with gasoline-biodiesel blend when using PPCI mode for either ambient intake pressure or increasing intake pressure at 0.12 MPa. The trend of graph shows that the increasing intake boosting also followed by increasing the NO_x emission for both diesel and gasoline-biodiesel blend. It is mean that the increasing of intake boosting has opposite function with 20% EGR. In this case, by using only 20% EGR rate, the NO_x emission of GCI engine fueled with gasoline-biodiesel blend is very low under 0.05 mg/kWh. However, increasing intake boosting 0.12 MPa, leads the deterioration on NO_x emission to be around 0.2 mg/kWh.

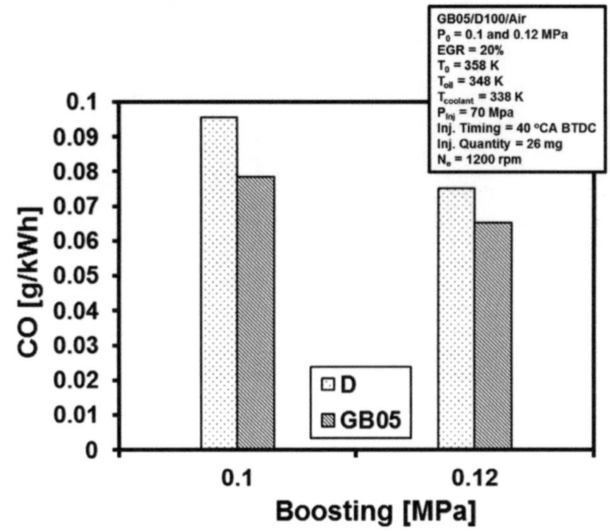

Figure 16.
Effect of boosting on CO emission of PPCI mode.

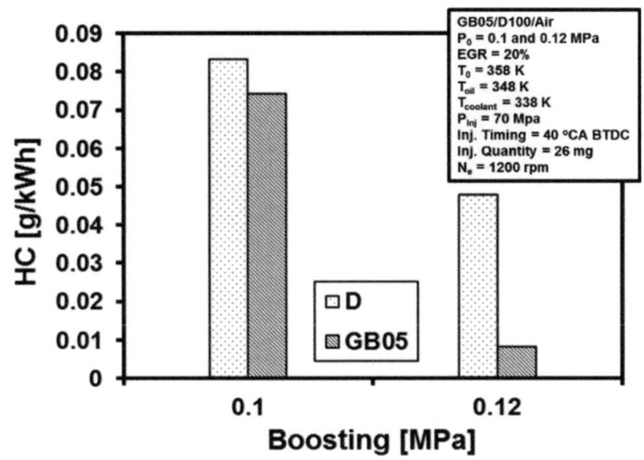

Figure 17.
Effect of boosting on HC emission of PPCI mode.

The effect of intake boosting on smoke emission of GCI engine with PPCI strategy can be seen in **Figure 19**. Smoke emission of GCI engine fueled with diesel fuel is very high almost 6 g/m^3 when running on PPCI mode by 20% of EGR rate and ambient pressure of intake boosting. While, in this condition smoke emission of GCI engine fueled with gasoline-biodiesel blend much lower than diesel fuel at around 1.5 g/m^3. Increasing intake boosting to be 0.12 MPa makes smoke emission of GCI engine fueled with diesel fuel decrease very significant around 3 g/m^3. However, the increasing of intake boosting to be 0.12 MPa for GCI engine fueled with gasoline-biodiesel caused the increasing of smoke emission, even though still lower than the emission of GCI engine fueled with diesel fuel which is to be around 2.5 g/m^3. If the point of view of GCI engine fueled with gasoline-biodiesel blend running on PPCI mode focused simultaneously on NOx emission and smoke emission, then it can be stated that the optimum effort to reduce both of emission parts is by using 20% EGR rate and 0.1 MPa intake boosting.

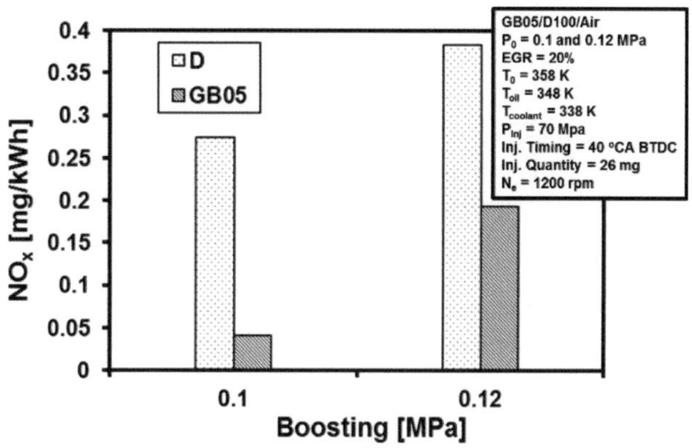

Figure 18.
Effect of boosting on NO_x emission of PPCI mode.

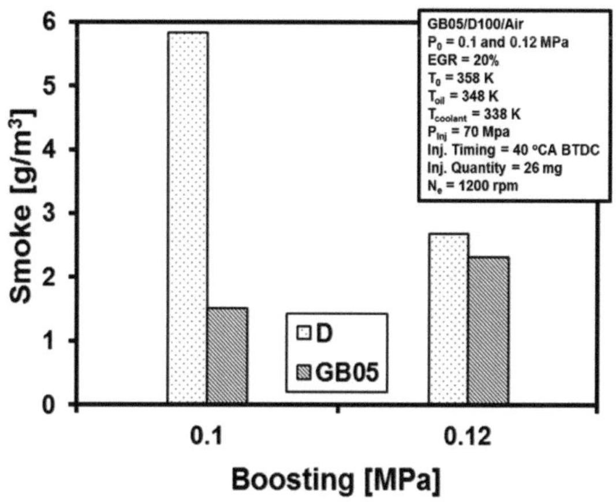

Figure 19.
Effect of boosting on smoke emission of PPCI mode.

4. Conclusions

The study on GCI engine was conducted in an experiment using biodiesel addition 5% into gasoline, compared to neat diesel with single injection (PPCI) strategy combined with the application of EGR and intake boosting in order to obtain high efficiency and low emission of GCI engine. The engine testing was set in the same of energy input that is injected fuel amount around 26 mg per cycle. Based on the results and comprehensive analysis, the following general conclusions may be drawn from this study:

Increasing EGR rate the value of indicated thermal efficiencies are decreased for both of diesel and gasoline-biodiesel blends. The highest 50% EGR rate for diesel fuel leads to a little increasing value of indicated thermal efficiency is compared with 20% of EGR rate. However, it caused the significant drop value of indicated thermal efficiency in case of gasoline-biodiesel blends fuel. By using diesel fuel, the 20% of EGR rate gives highest NO_x emission. Even though, when 50% EGR was applied the NO_x emission will also decreasing. The utilization of EGR gives effect on the drop of NO_x emission value for gasoline-biodiesel blend much lower than diesel fuel. However, there are no effects of EGR rate variations on NO_x emission of GCI engine fueled with gasoline-biodiesel blends. Smoke emission of GCI engine fueled with diesel in the high level for all variation of EGR rate, even when the rate increased. However, the smoke emission of GCI engine using gasoline-biodiesel blends obtains its lowest value when EGR rates at 20%.

The indicated thermal efficiency of GCI engine fueled with diesel fuel increased due to the increasing of intake boosting. Similarly, for GCI engine fueled with gasoline-biodiesel blend, the indicated thermal efficiency was also increased when the intake boosting increased to be 0.12 MPa. The NO_x emission of GCI engine fueled with diesel is higher than GCI engine fueled with gasoline-biodiesel blend when using PPCI mode for either ambient intake pressure or increasing intake pressure at 0.12 MPa. The increasing intake boosting also followed by increasing NO_x emission for both diesel and gasoline-biodiesel blend. Increasing intake boosting to be 0.12 MPa makes smoke emission of GCI engine fueled with diesel fuel decrease very significant around 3 g/m^3. However, the increasing of intake boosting to be 0.12 MPa for GCI engine fueled with gasoline-biodiesel caused the increasing of smoke emission, even though still lower than the emission of GCI engine fueled with diesel fuel.

Acknowledgements

This research was supported by the University of Ulsan, Korea. Yanuandri Putrasari acknowledges the support from the Indonesian Institute of Sciences (LIPI) and the Ministry of Research and Technology/National Research and Innovation Agency of Republic Indonesia (RISTEK-BRIN).

Author contribution

All authors contributed equally as the main contributor of this chapter. All authors read and approved the final version of this chapter.

Conflict of interest

The authors declare no conflict of interest.

Author details

Yanuandri Putrasari[1] and Ocktaeck Lim[2*]

1 Research Centre for Electrical Power and Mechatronics – Indonesian Institute of Sciences (LIPI), Bandung, Indonesia

2 School of Mechanical Engineering, University of Ulsan, Ulsan, South Korea

*Address all correspondence to: otlim@ulsan.ac.kr

References

[1] Dec JE. Advanced compression-ignition engines - Understanding the in-cylinder processes. Proc Combust Inst 2009;32 II:2727–42. doi:10.1016/j.proci.2008.08.008.

[2] Eng JA. Characterization of Pressure Waves in HCCI Combustion Reprinted From : Homogeneous Charge Compression Ignition Engines. SAE Tech Pap 2002;1:15. doi:10.4271/2002-01-2859.

[3] Das P, Subbarao PM V, Subrahmanyam JP. Control of combustion process in an HCCI-DI combustion engine using dual injection strategy with EGR. Fuel 2015;159:580–9. doi:10.1016/j.fuel.2015.07.009.

[4] Han D, Ickes AM, Bohac S V, Huang Z, Assanis DN. HC and CO emissions of premixed low-temperature combustion fueled by blends of diesel and gasoline. Fuel 2012;99:13–9. doi:10.1016/j.fuel.2012.04.010.

[5] Yang B, Li S, Zheng Z, Yao M, Cheng W. A comparative study on different dual-fuel combustion modes fuelled with gasoline and diesel. SAE Tech Pap 2012;2012–01–06. doi:10.4271/2012-01-0694.

[6] Kweon C, Foster D, Shibata G. The effects of oxygenate and gasoline-diesel fuel blends on diesel engine emissions. SAE Tech Pap 2000;2000–01–11. doi:https://doi.org/10.4271/2000-01-1173.

[7] Iannuzzi SE, Valentino G. Comparative behavior of gasoline-diesel / butanol-diesel blends and injection strategy management on performance and emissions of a light duty diesel engine. Energy 2014;71:321–31. doi:10.1016/j.energy.2014.04.065.

[8] Prikhodko VY, Curran SJ, Barone TL, Lewis S a, Storey JM, Cho K, et al. Emission Characteristics of a Diesel Engine Operating with In-Cylinder Gasoline and Diesel Fuel Blending. SAE Int 2010;2266:946–55.

[9] Leermakers CAJ, Van den Berge B, Luijten CCM, Somers LMT, de Goey LPH, Albrecht BA. Gasoline-Diesel Dual Fuel: Effect of Injection Timing and Fuel Balance. SAE Pap 2011;2011–01–24. doi:10.4271/2011-01-2437.

[10] Liu H, Wang Z, Wang J, He X. Improvement of emission characteristics and thermal ef fi ciency in diesel engines by fueling gasoline / diesel / PODEn blends. Energy 2016;97:105–12. doi:10.1016/j.energy.2015.12.110.

[11] Bae C, Kim J. Alternative fuels for internal combustion engines. Proc Combust Inst 2017;36:3389–413. doi:10.1016/j.proci.2016.09.009.

[12] Ra Y, Yun JE, Reitz RD. Numerical Parametric Study of Diesel Engine Operation with Gasoline. Combust Sci Technol 2009;181:2:350–78. doi:10.1080/00102200802504665.

[13] Rose KD, Ariztegui J, Cracknell RF, Dubois T, Hamje HDC, Pellegrini L, et al. Exploring a gasoline compression ignition (GCI) engine concept. SAE Int 2013;2013–01–09:1–54. doi:10.4271/2013-01-0911.

[14] Badra J, Viollet Y, Elwardany A, Im HG, Chang J. Physical and chemical effects of low octane gasoline fuels on compression ignition combustion. Appl Energy 2016;183:1197–208. doi:10.1016/j.apenergy.2016.09.060.

[15] Kim K, Kim D, Jung Y, Bae C. Spray and combustion characteristics of gasoline and diesel in a direct injection compression ignition engine. Fuel 2013;109:616–26. doi:10.1016/j.fuel.2013.02.060.

[16] Yang B, Yao M, Zheng Z, Yue L. Experimental Investigation of Injection Strategies on Low Temperature Combustion Fuelled with Gasoline in a Compression Ignition Engine 2015;2015.

[17] Kim D, Bae C. Application of double-injection strategy on gasoline compression ignition engine under low load condition. Fuel 2017;203:792–801. doi:10.1016/j.fuel.2017.04.107.

[18] Kodavasal J, Kolodziej CP, Ciatti SA. Effects of injection parameters , boost , and swirl ratio on gasoline compression ignition operation at idle and low-load conditions. Int J Engine Res 2016:1–13. doi:10.1177/1468087416675709.

[19] Loeper P, Ra Y, Foster D, Ghandhi J. Experimental and computational assessment of inlet swirl effects on a gasoline compression ignition (GCI) light-duty diesel engine. SAE 2014 World Congr Exhib 2014;1. doi:10.4271/2014-01-1299.

[20] Kalghatgi GT, Risberg P, Ångström H. Advantages of Fuels with High Resistance to Auto-ignition in Late-injection, Low-temperature, Compression Ignition Combustion. SAE Int 2006:SAE 2006-01-3385. doi:10.4271/2006-01-3385.

[21] Kalghatgi GT, Risberg P, Angstrom H-E. Partially Pre-Mixed Auto-Ignition of Gasoline to Attain Low Smoke and Low NOx at High Load in a Compression Ignition Engine and Comparison with a Diesel Fuel. SAE Tech Pap 2007;2007–01–00. doi:10.4271/2007-01-0006.

[22] Yang H, Shuai S, Wang Z, Wang J. Fuel octane effects on gasoline multiple premixed compression ignition (MPCI) mode. Fuel 2013;103:373–9. doi:10.1016/j.fuel.2012.05.016.

[23] Wang B, Wang Z, Shuai S, Xu H. Combustion and emission characteristics of Multiple Premixed Compression Ignition (MPCI) mode fuelled with different low octane gasolines. Appl Energy 2015;160:769–76. doi:10.1016/j.apenergy.2015.01.115.

[24] Kodavasal J, Kolodziej CP, Ciatti SA, Sibendu S. Computational Fluid Dynamics Simulation of Gasoline Compression Ignition 2017;137:1–13. doi:10.1115/1.4029963.

[25] Won HW, Peters N, Pitsch H, Tait N, Kalghatgi G. Partially premixed combustion of gasoline type fuels using larger size nozzle and higher compression ratio in a diesel engine. SAE Tech Pap 2013;11. doi:10.4271/2013-01-2539.

[26] Tesfa B, Mishra R, Zhang C, Gu F, Ball AD. Combustion and performance characteristics of CI (compression ignition) engine running with biodiesel. Energy 2013;51:101–15. doi:10.1016/j.energy.2013.01.010.

[27] Cordiner S, Mulone V, Nobile M, Rocco V. Impact of biodiesel fuel on engine emissions and Aftertreatment System operation. Appl Energy 2016;164:972–83. doi:10.1016/j.apenergy.2015.07.001.

[28] Rakopoulos CD, Rakopoulos DC, Hountalas DT, Giakoumis EG, Andritsakis EC. Performance and emissions of bus engine using blends of diesel fuel with bio-diesel of sunflower or cottonseed oils derived from Greek feedstock. Fuel 2008;87:147–57. doi:10.1016/j.fuel.2007.04.011.

[29] Wang Z, Li L, Wang J, Reitz RD. Effect of biodiesel saturation on soot formation in diesel engines. Fuel 2016;175:240–8. doi:10.1016/j.fuel.2016.02.048.

[30] Adams CA, Loeper P, Krieger R, Andrie MJ, Foster DE. Effects of biodiesel-gasoline blends on gasoline direct-injection compression ignition

(GCI) combustion. Fuel 2013;111:784–90. doi:10.1016/j.fuel.2013.04.074.

[31] Putrasari Y, Lim O. A study on combustion and emission of GCI engines fueled with gasoline-biodiesel blends. Fuel 2017;189:141–54. doi: 10.1016/j.fuel.2016.10.076.

[32] Putrasari Y, LIM O. A study of a GCI engine fueled with gasoline-biodiesel blends under pilot and main injection strategies. Fuel 2018;221:269–82. doi: 10.1016/j.fuel.2018.01.063.

[33] Yang B, Li S, Zheng Z, Yao M, Cheng W. A Comparative Study on Different Dual-Fuel Combustion Modes Fuelled with Gasoline and Diesel. SAE Int 2012;694. doi:10.4271/2012-01-0694.

[34] Liu J, Shang H, Wang H, Zheng Z, Wang Q, Xue Z, et al. Investigation on partially premixed combustion fueled with gasoline and PODE blends in a multi-cylinder heavy-duty diesel engine. Fuel 2017;193:101–11. doi: 10.1016/j.fuel.2016.12.045.

[35] Cairns A, Blaxill H. The Effects of Combined Internal and External Exhaust Gas Recirculation on Gasoline Controlled Auto-Ignition 2005;2005. doi:10.4271/2005-01-0133.

[36] Zhao H, Peng Z, Williams J, Ladommatos N. Understanding the Effects of Recycled Burnt Gases on the Controlled Autoignition (CAI) Combustion in Four-Stroke Gasoline Engines 2001. doi:10.4271/2001-01-3607.

[37] Olsson J-O, Tunestål P, Ulfvik J, Johansson B. The effect of cooled EGR on emissions and performance of a turbocharged HCCI engine. Soc Automot Eng 2003;2003:21–38. doi: 10.4271/2003-01-0743.

[38] Yao M, Chen Z, Zheng Z, Zhang B, Xing Y. Effect of EGR on HCCI Combustion fuelled with Dimethyl Ether (DME) and Methanol Dual-Fuels. SAE Tech Pap 2005;2005–01–37. doi:10.4271/2005-01-3730.

[39] Sjöberg M, Dec JE, Hwang W. Thermodynamic and Chemical Effects of EGR and Its Constituents on HCCI Autoignition. SAE Tech Pap 2007;2007–01–02:776–90. doi:10.4271/2007-01-0207.

[40] Christensen M, Johansson B. Supercharged Homogeneous Charge Compression Ignition (HCCI) with Exhaust Gas Recirculation and Pilot Fuel. SAE Tech Pap 2000:SAE 2000-01-1835. doi:10.4271/2000-01-1835.

[41] Sjöberg M, Dec JE. EGR and Intake Boost for Managing HCCI Low-Temperature Heat Release over Wide Ranges of Engine Speed 2007:776–90. doi:10.4271/2007-01-0051.

[42] Saxena S, Bedoya ID. Fundamental phenomena affecting low temperature combustion and HCCI engines, high load limits and strategies for extending these limits. Prog Energy Combust Sci 2013;39:457–88. doi:10.1016/j.pecs.2013.05.002 Review.